Innovation in Seed Potato Systems
in Eastern Africa

Royal Tropical Institute (KIT)
Development Policy and Practice
PO Box 95001
1090 HA Amsterdam
The Netherlands
www.kit.nl

International Potato Center (CIP)
PO Box 1558
Lima, Peru
www.cipotato.org

Wageningen University and Research Centre
(Wageningen UR)
PO Box 9101
6700 HB Wageningen
The Netherlands
www.wur.nl

Thesis supervisors
Prof. dr. ir. P.C. Struik
Professor of Crop Physiology
Wageningen University

Prof. dr. ir. C. Leeuwis
Professor of Communication and Innovation Studies
Wageningen University

Other thesis committee members
Prof. dr. P. Richards
Wageningen University
Prof. dr. ir. M.K. van Ittersum
Wageningen University
Prof. dr. ir. A.J. Haverkort
Wageningen University and Research Centre
Dr. A. Hall
LINK Ltd.; UNU-MERIT, Maastricht;
Open University, Milton Keynes, UK

This research was conducted under the auspices of the
C.T. de Wit Graduate School for Production Ecology
and Resource Conservation (PE&RC).

Important components of this research have been
supported by project grants from the International
Fund for Agricultural Development (IFAD), the OPEC
Fund for International Development, and the United
States Agency for International Development (USAID),
by CIP's core funds between 2004 and 2007 and by the
Directorate-General for International Cooperation
(DGIS) of the Dutch Ministry of Foreign Affairs .

The views and opinions expressed in this publication
are those of the authors and do not necessarily reflect
the official policy or position of the three co-publishing
organisations or the funders.

Publication and distribution by KIT Publishers
P.O. Box 95001
1090 HA Amsterdam
The Netherlands
Telephone +31 (0)20 568 8272
Fax +31 (0)20 568 8286
www.kitpublishers.nl
Email: publishers@kit.nl

© 2012 KIT Publishers, Amsterdam, The Netherlands

Cover and design
   Grafisch Ontwerpbureau Agaatsz bNO, Meppel,
   The Netherlands

Printing
   Drukkerij Bariet, Steenwijk, The Netherlands

ISBN: 978-94-6022-407-2

Innovation in Seed Potato Systems in Eastern Africa

Peter R. Gildemacher

Thesis
submitted in fulfilment of the
requirements for the degree of doctor
at Wageningen University
by the authority of the Rector Magnificus
Prof. dr. M.J. Kropff,
in the presence of the
Thesis Committee appointed by the Academic Board
to be defended in public
on Wednesday 20 June 2012
at 4 p.m. in the Aula.

Royal Tropical Institute

Voor Anje

Table of contents

Abstract			12
1	**General Introduction**		**15**
	1.1	Background	15
		1.1.1 Sub-Sahara Africa	15
		1.1.2 Role of agriculture in Sub-Sahara Africa	15
		1.1.3 Potato in Sub-Sahara Africa	15
		1.1.4 Problem statement	18
	1.2	Objectives and approach	19
		1.2.1 Identification of entry points for potato system improvement	20
		1.2.2 Zooming in on seed potato quality improvement	20
		1.2.3 Improving the quality self-supply seed potatoes through positive selection	21
		1.2.4 Investigating the mechanisms behind the effects of positive selection on seed potato quality	21
		1.2.5 Analysing the role of agricultural research in innovation	22
		1.2.6 Implication for seed potato system improvement and conclusions on the role of research	22
	References		23
2	**Improving potato production in Kenya, Uganda and Ethiopia: a system diagnosis**		**25**
	Abstract		26
	2.1	Introduction	26
	2.2	Methodology	27
		2.2.1 Potato practices and technology survey	27
		2.2.2 Knowledge and information survey	28
		2.2.3 Stakeholder meetings to assess potato related innovation system	29
	2.3	Potato production system characterization	30
		2.3.1 Trends in potato production and productivity	30
		2.3.2 Potato production seasons	32
		2.3.3 Importance of the potato crop in the farming system	32
		2.3.4 Use of agricultural inputs	34
		2.3.5 Seed potato source and renewal	35
	2.4	Potato marketing system characterization	36
		2.4.1 Home consumption versus marketing	36
		2.4.2 Potato marketing channels	36
	2.5	Potato production economics	37
		2.5.1 Profitability of potato production	37
		2.5.2 Return on cash investment	38
		2.5.3 Return to family labour	38

2.6		Identification of technical yield reducing factors and innovation priorities	39
	2.6.1	Improving productivity	39
	2.6.2	Seed potato quality management	40
	2.6.3	Bacterial wilt control	40
	2.6.4	Late blight control	41
	2.6.5	Soil fertility management	42
2.7		Potato innovation system	42
	2.7.1	Potato marketing	43
	2.7.2	Knowledge development and information exchange	43
	2.7.3	Quantification of information sources of potato producers	44
	2.7.4	Important actors in the potato related innovation systems	47
2.8		Potato innovation system constraints identified	49
	2.8.1	International potato center	49
	2.8.2	National research organizations	50
	2.8.3	Agricultural extension actors	51
	2.8.4	Potato growers	51
	2.8.5	Private sector	52
2.9		Discussion and conclusions	53
	2.9.1	Potato production and marketing	53
	2.9.2	Potato innovation system	53
References			55

3 Seed potato systems in East Africa: description and opportunities for improvement   **59**

Abstract			60
3.1		Introduction	60
3.2		Materials and methods	61
	3.2.1	Quantification of the importance of seed borne diseases	61
	3.2.2	Potato farming practices survey	62
3.3		Results	62
	3.3.1	Potato virus survey in Kenya	62
	3.3.2	Bacterial wilt survey in Kenya	63
	3.3.3	Potato farming practices survey	63
	3.3.4	Seed potato sources	64
	3.3.5	Seed renewal period	64
	3.3.6	Seed potato management practices by ware potato producers	64
	3.3.7	Specialized seed potato multiplication	66
	3.3.8	Estimated volumes of seed potatoes in the specialized seed potato chain	66
	3.3.9	Seed potato economics	68
3.4		Discussion	69
	3.4.1	Potato disease levels	69
	3.4.2	Seed potato management by ware producers	69
	3.4.3	Seed potato economics	70
	3.4.4	Improving seed quality in the local seed potato chain	70
	3.4.5	Improving efficiency in the specialized chain	71
References			72

4 Seed potato quality improvement through positive selection by smallholder farmers in Kenya — 75

Abstract — 76
4.1 Introduction — 76
    4.1.1 Seed potato systems in Kenya — 76
    4.1.2 Positive seed potato selection — 78
4.2 Materials and Methods — 79
    4.2.1 Training methodology — 79
    4.2.2 Data collection and analysis — 80
4.3 Results — 81
    4.3.1 Experimental results — 81
    4.3.2 Economic analysis — 84
4.4 Discussion — 85
References — 88

5 Improving seed health and seed performance by positive selection in three Kenyan potato varieties — 91

Abstract — 92
5.1 Introduction — 92
5.2 Materials and methods — 93
    5.2.1 Source fields and starting material — 93
    5.2.2 Replicated farmer managed on-farm trials — 93
    5.2.3 Virus infection level testing — 94
    5.2.4 Replicated on-station factorial fertilizer seed source trials — 94
5.3 Results — 95
5.4 Discussion — 97
    5.4.1 Positive selection results in yield increase compared with farmer selection — 97
    5.4.2 Positive selection reduces virus infection compared with farmer practice — 98
    5.4.3 The relationship between yield increase and virus reduction — 98
    5.4.4 Possible additional mechanisms contributing to the effect of positive selection — 99
    5.4.5 Effect of soil fertility on effectiveness of positive selection — 101
    5.4.6 Remaining research questions — 101
    5.4.7 Consequences of the research findings — 102
References — 103

6 Dissecting a successful research-led innovation process: the case of positive selection in seed potato production in Kenya — 107

Abstract — 108
6.1 Introduction — 108
6.2 Positive seed potato selection — 110
    6.2.1 Seed potato systems in Kenya — 110
    6.2.2 Positive seed potato selection — 110

		6.2.3	Positive selection programme in Kenya	111
		6.2.4	Technical results of positive selection by smallholder producers	112
		6.2.5	Initial adoption of positive selection by trained farmers	112
		6.2.6	Discussion of the results	113
		6.2.7	Scaling-up	114
		6.2.8	A case of successful innovation	114
	6.3	Specific characteristics of the positive selection intervention		114
		6.3.1	Decision making leading to the initiative	116
		6.3.2	Project partnership configuration	117
		6.3.3	Funding	118
		6.3.4	Freedom and flexibility	119
		6.3.5	Type of technology	119
		6.3.6	Approach to validation and adaptation	120
		6.3.7	Characteristics of the technology	121
		6.3.8	Training methodology	121
	6.4	Analysis and discussion		122
		6.4.1	Role of the researchers	122
		6.4.2	Research responsibilities in the training	122
		6.4.3	Goal and phasing of activities	123
		6.4.4	Enabling or hindering conditions within the research organization	124
		6.4.5	Enabling or hindering external conditions	125
		6.4.6	Relevance of the case to other innovation trajectories	126
	6.5	Conclusion and implications for research and development		126
	References			128
7	**General discussion**			**133**
	7.1	Introduction		134
	7.2	Potato system diagnosis		135
	7.3	Seed potato system diagnosis		136
		7.3.1	Potato diseases	136
		7.3.2	Seed potato stock replacement in East Africa	137
		7.3.3	Seed potato quality management by ware potato producers	137
		7.3.4	Improving the specialized seed potato chain	139
		7.3.5	Improving the non-specialized seed potato production	139
	7.4	Positive seed potato selection		141
		7.4.1	What makes positive selection attractive for smallholder producers?	142
		7.4.2	Limitations of positive selection	143
	7.5	Consequences for the future of seed potato systems in Sub-Sahara Africa		144
		7.5.1	Consequences for seed potato system development interventions	144
		7.5.2	Consequences for further seed potato systems research	144
	7.6	Lessons on the role of agricultural research in accelerating innovation		145
		7.6.1	Some innovation concepts applied to the potato system research	146
		7.6.2	The role of agricultural research in technical innovation	147
		7.6.3	Opportunity assessment to identify entry points for innovation	149
		7.6.4	The pilot innovation process under realistic circumstances	149

7.6.5	Researcher controlled testing	151
7.6.6	Scaling-up for impact	151
7.6.7	Feedback	154
7.6.8	Additional considerations regarding the process of deliberate innovation	154
7.6.9	Consequences for agricultural research	155
References		157

Summary	165
Samenvatting	171
Acknowledgements	177
Publication list	179
About the author	182
PE&RC PhD Education Certificate	183

Abstract

Gildemacher, Peter R., 2012. Innovation in Seed Potato Systems in Eastern Africa. Thesis, Wageningen University, Wageningen, NL, with references and summaries in English and Dutch. ISBN 978-94-6173-310-8, 184 pp.

Potato (*Solanum tuberosum* L.) is a crop with a high potential to contribute to poverty reduction in Eastern Africa through income increase and improved food security. It is largely grown by smallholders, has a high production per hectare, stable prices and a steadily growing demand. Average yields in Eastern Africa of 10.5 t/ha are much below the world average and yields observed in the fields of better performing smallholders. There is both a need and a potential for increased potato productivity.

A diagnosis of the potato systems of Kenya, Uganda and Ethiopia identified integrated management of bacterial wilt (*Ralstonia solanacearum*) and late blight (*Phytophthora infestans*), soil fertility management, and improving seed potato quality as technology-based opportunities for innovation. Improvement of potato supply chains and the knowledge exchange in the sector were identified as systemic opportunities for improvement. Analysis of the seed potato system confirms that both virus diseases and bacterial wilt are likely contributors to the low yields. In only 3% of seed potato tubers sold in rural markets was free of PVY, PLRV, PVX and PVA. *Ralstonia solanacearum* was found in 74% of potato farms sampled in Kenya.

Less than 5% of the seed potatoes used are sourced from specialized multipliers (specialized chain). Farmers rely on seed potatoes from neighbours and farm-saved seed potatoes (local chain). This often makes economic sense in the absence of affordable high quality seed potatoes and limited market security. Seed potato system interventions need to tackle the local and specialized chain simultaneously. Private investment in specialized multiplication could stimulate the production of affordable high-quality seed potatoes. For local chain improvement training on seed quality maintenance and on bacterial wilt and virus management is needed. Research into the mitigation of yield-reducing effects of indiscriminate seed potato recycling, such as research on virus resistance and positive selection deserves attention.

Positive selection, the selection of healthy looking mother plants for the production of seed potatoes, can contribute to improving quality management in the local chain. In farmer managed trials in Kenya it gave an average yield increase of 34% which corresponded to a 284 Euro profit increase per ha. It requires no cash and only 4 man-days per hectare and is an important alternative and complementary technology to regular seed replacement.

In 18 replicated trials it was shown that positive selection lowered the incidence of PLRV, PVY and PVX with 35%, 35% and 39%, respectively, and increased yields irrespective of the agro-ecology, crop management, soil fertility, variety, and quality of the starting material with an average 30%, compared to current farmer practice. Regression analysis showed that this reduction in virus incidence contributed to the higher yields, but did not fully account for the effect. Probably other, not tested, virus diseases and other seed borne diseases also played a role. It can be concluded that positive selection can benefit all smallholder potato producers who select seed potatoes from their own fields, and should thus be incorporated routinely in agricultural extension efforts.

In retrospect the research trajectory can be considered a successful contribution of agricultural research to innovation. It shows that it is worthwhile to search for opportunities for incremental innovation that do not require institutional change and that these opportunities can be of a surprising simple nature, and based on old technology. Essential for researchers to contribute to innovation is room to manoeuvre and opportunity to immerse in practical collaborative partnerships with practitioners. Most importantly, innovation needs to be made a central objective, rather than research results, and the mandate of research needs to be broadened and allow for the active engagement in training, communication and scaling-up.

Keywords: Potato, positive selection, viruses, seed potato systems, East Africa, Kenya, Uganda, Ethiopia

1 General introduction

Seed is the starting point of plant life, and hence the most fundamental input of agriculture. A seed system that assures the availability of the desired quality of seed to the producer at the right time is indispensable for his farming enterprise. In the case of the potato crop, the seed most commonly used is strictly speaking no seed, but a tuber. The constraints and opportunities in seed potato systems in East Africa are of a combined social, economic and technical nature. This thesis presents the results of an interdisciplinary research and development programme that aimed at innovation of seed potato systems in East Africa.

1.1 Background

1.1.1 Sub-Sahara Africa

African countries have over the last two decades achieved relatively high economic growth rates, with a continent-wide average of 6.1% per year between 2003 and 2007 (UNECA, 2010). Part of the effects of economic growth on the prosperity of the population have been offset by an annual growth rate in population on the African continent of 2.3% (UNFPA, 2011). In spite of the high population growth, the continent as a whole has experienced an annual increase in Gross Domestic Product (GDP) per capita since 1995 (UNECA, 2010). During the recent economic crisis there was a drop in the economic growth rate in Sub-Sahara Africa, but Sub-Sahara Africa is expected to realise economic growth above 5% in both 2011 and 2012 (IMF, 2011; UNECA, 2011).

At the same time, however, Sub-Sahara Africa still harbours 33 of the total of 49 countries that fall in the category of Least Developed Countries (LDC) (UNCTAD, 2011). In addition, no drop in population growth is predicted in Sub-Sahara Africa for the coming decades. The population of the African continent is expected to more than triple this century from 1 billion in 2011 to an estimated 3.6 billion in 2100 (UNFPA, 2011). Nigeria is expected to become the third most populous nation by 2100 with 730 million inhabitants and 10 out of the 20 most populous nations are expected to be in Sub-Sahara Africa, against 3 in 2010 (UN-ESA, 2011). Assuring further growth of GDP per capita, while assuring access to affordable food for the fast growing population is the main challenge for Sub-Sahara Africa. The food crisis in 2008 came as a shock but food prices dropped after that. The FAO (2011) warned, however, to have an eye for the underlying longer term trend of increasing food prices since 2002, which is expected to continue (IFAD, 2010). Furthermore it is expected that price volatility of food will increase further (FAO, 2011). This price volatility makes both smallholder producers and poor consumers increasingly vulnerable to poverty (FAO, 2011). The recent food crisis also showed that grain prices were much more volatile than prices of root and tuber crops (Cromme et al., 2010).

1.1.2 Role of agriculture in Sub-Sahara Africa

In the short run high food prices have a negative impact on food security, in urban centres, but also in rural areas. For the longer term however, these higher prices also provide incentives for long-term investment in agriculture (FAO, 2011). Further investment in agriculture is critical to sustainable long-term food security. In addition, agriculture provides promising opportunities for rural economic development in the majority of Sub-Sahara African countries (IFAD, 2010; UNECA, 2011). From the point of view of poverty reduction, agricultural based economic growth has a stronger effect than non-agricultural growth. As such agricultural development efforts can provide high payoffs towards the achievement of the millennium development goals and beyond (Worldbank, 2007b). In comparison to Asia, the green revolution did not deliver the desired productivity increases in Africa. This can largely be attributed to a poor fit between the technological packages promoted by the green revolution and the African socio-economic and agro-ecological reality. Amongst other factors the heterogeneous nature of its largely rainfed agriculture, low soil fertility, poorly functioning markets and infrastructure, and low public spending on agriculture contributed to the limited success of the green revolution technologies. There are, however, signals that Sub-Sahara Africa is turning a corner, and productivity increase through intensification is happening (Worldbank, 2007b).

1.1.3 Potato in Sub-Sahara Africa

Potato (*Solanum tuberosum* L.) is a crop with a high potential to contribute to poverty reduction in Sub-Sahara Africa. The potato crop can contribute to improving food security. First and foremost its productivity in terms of energy produced per hectare per day is the highest of all major arable crops, and almost double that of wheat and rice (Scott et al., 2000). This is an essential asset of the crop, considering the increasing population pressure necessitating intensification of food production. In addition the growing season of the potato crop is short. Especially in rainfed systems this is of essence, as it makes potato one of the first crops that can be harvested after the onset of the rainy season. In situations of food insecurity this makes potato an important 'hunger breaking' crop to assure staple food before grains can be harvested. The crop is well appreciated for this in for example Ethiopia. In addition it was observed that during the food crisis potato prices, as the prices of other root and tuber crops, fluctuated much less than prices of major cereals. As ware potatoes are largely locally and nationally traded, prices reflect local demand and supply, and are less susceptible to international food price speculation than cereals (Cromme et al., 2010).

Besides its potential to contribute to food security, potato provides opportunities for rural economic development. Scott et al. (2000) demonstrated that there is a steadily growing demand for potatoes in Sub-Sahara Africa, a trend which is expected to continue. Current potato consumption levels per capita in Sub-Sahara Africa are still rather low, but growing steadily. This is the result of fast population growth and even faster urbanisation. In addition food patterns are changing in conjunction with urbanisation and a change in income, and as a result the demand for potatoes as a staple and snack food is growing. Scott et al. (2000) shows that potato consumption rises with a rise in income. The prospect of a steady rise in potato consumption provides opportunities for production increase of potato without an erosion of farm gate prices. Moreover, it provides opportunities for creating added value in the snack food chain.

Potato is a commercial commodity in most of Sub-Sahara Africa, and its market is growing. Potatoes require cannot be grown everywhere. When the temperature is high during critical stages of tuber formation potato plants tend to favour above ground growth of the foliage over tuber formation and growth, resulting in very poor yields (Ewing, 1981). Favourable conditions for potato production are found throughout Sub-Sahara Africa, for example in the East African highlands, but also in the Sahelian zone during the Harmattan season and in Southern Africa during the cold season. The commodity is demanded also outside these suitable agro-ecologies, making it a widely traded commodity over short distances. An added advantage is that the market for potato is largely national, making the trade dynamic and quick, with in most cases on the spot payment at the farmgate or collection point to potato producers.

A fourth factor contributing to the potential of potato for economic growth and poverty reduction is that it is largely grown by smallholder producers, thus increasing the chances that improvements in the potato sector benefit directly poor rural households. As potatoes are relatively easy to market, but at the same time form a staple food crop, improved potato production has the potential to simultaneously increase cash income of rural producer households and to directly contribute to increased household food security.

The trends over the last two decades demonstrate that the production of potatoes in Eastern Africa has almost quadrupled, which can largely be explained by an increase in potato area (Figure 1.1). Southern Asia showed a similar trend while potato production in South America and Western Europe have shown modest growth or have been stable, respectively.

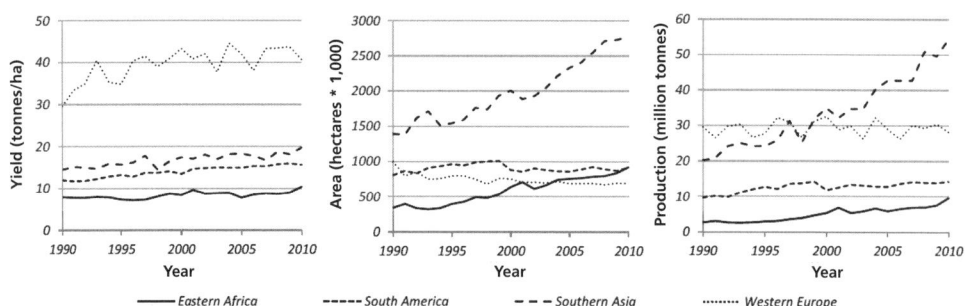

Figure 1.1: Trends in potato production in Eastern Africa, South America and Western Europe compared.
Source: FAOSTAT, foa.faostat.org, January 2012
Eastern Africa = Burundi; Comoros; Eritrea; Ethiopia; Kenya; Madagascar; Malawi; Mauritius; Mozambique; Réunion; Rwanda; Uganda; United Republic of Tanzania; Zambia; Zimbabwe
South America = Argentina; Bolivia; Brazil; Chile; Colombia; Ecuador; Paraguay; Peru; Uruguay; Venezuela
South Asia = Afghanistan; Bangladesh; Bhutan; India; Iran; Nepal; Pakistan; Sri Lanka
Western Europe = Austria; Belgium; France; Germany; Luxembourg; Netherlands; Switzerland

1.1.4 Problem statement

The average current potato yields in Eastern Africa of 8-10 tonnes are well below the average of other sub-regions (Figure 1.1), and also much below the yields that are observed in the fields of better performing smallholders. Over the last decades the growing demand for potatoes in Eastern Africa has been met by an increase in area under production, while yields have been rather stable at a low level of 8-10 tonnes per hectare. The expected continuing increase in demand can no longer be met by further increases in areas under potato cultivation. A further increase in area under potato cultivation is not the solution for the future, as it will further shorten rotations which brings high risks of build-up of soil-borne diseases.

In this thesis the general question is posed how smallholder potato productivity and profitability can be increased in Kenya, Uganda and Ethiopia to respond to the increasing market demand. As the underlying reasons for low average productivity of potatoes are of a combined socio-economic and technical nature, attempts to address them will require to be interdisciplinary, and address the seed potato system as such, as well as the related innovation support system and socio-economic environment.

1.2 Objectives and approach

In this thesis the results of a research programme are presented that responded to the above problem statement. The research programme developed organically over time, in response to the incrementing sights gained. The insights of one component of the research programme fed the research questions and design of the next component. Figure 1.2 provides an overview of the chronology and relation between the different components of the research programme.

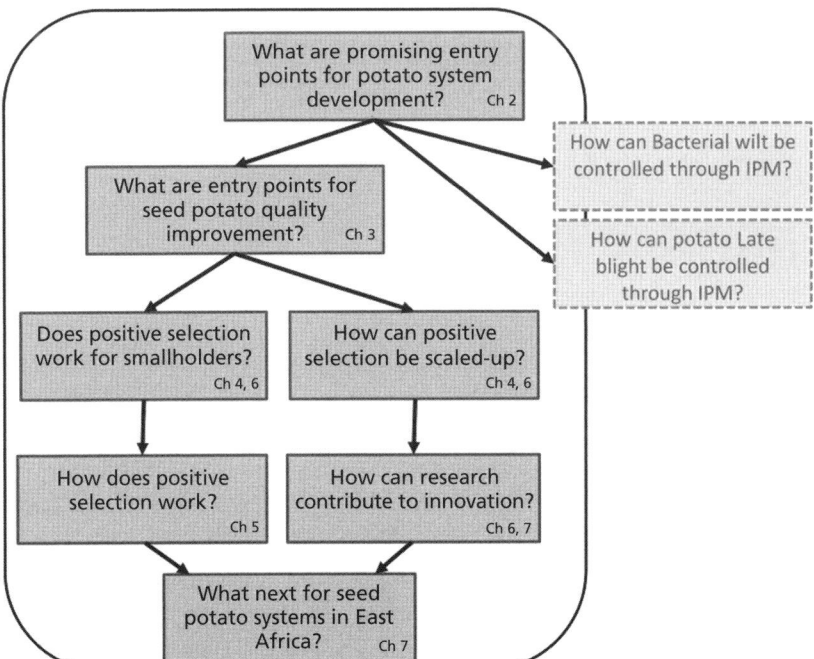

Figure 1.2: The thesis research components and their interrelations.
IPM = integrated pest management

1.2.1 Identification of entry points for potato system improvement

How can the potato sector be improved to the benefit of food security and smallholder potato producer livelihood in Kenya, Uganda and Ethiopia? This can be considered as the basic question posed at the onset of the research programme. To answer this question a diagnosis of the potato system in Kenya, Uganda and Ethiopia was made. The system diagnosis in the three countries was based on a combination of quantitative data collection through surveys and qualitative data collection through potato stakeholder meetings and interviews with key informants. The study focussed on answering the following, interrelated research questions:
1. Are there constraints in the knowledge and information flow in the potato system in the three countries?
2. What are production related constraints that contribute to the low potato yields in the region?
3. What are promising entry points for potato sector improvement?

Chapter 2 of the thesis presents the results of the system diagnosis. The 'system failure framework' (Woolthuis et al., 2005) is used to analyse the data and to identify key technical, socio-economic and system constraints hampering potato sector development. From these constraints promising entry points for intervention were identified. Integrated management of bacterial wilt (*Ralstonia solanacearum*) and late blight (*Phytophthora infestans*), soil fertility management and improving seed potato quality were identified as technology based opportunities for innovation. Improvement of potato supply chains and improving the knowledge exchange in the sector were identified as more systemic opportunities for potato sector improvement.

In response to the identified opportunities further research trajectories were initiated in the field of seed quality improvement, Integrated Pest Management (IPM) of bacterial wilt and late blight. In this thesis the further research related to seed systems is presented, while further research on late blight and bacterial wilt IPM, carried out by the author, is not discussed in this thesis.

1.2.2 Zooming in on seed potato quality improvement

A specific analysis of the seed potato system in Kenya, Uganda and Ethiopia was executed to investigate whether innovative approaches could be identified to seed potato system innovation. This was deemed necessary in light of the perceived limited successes of more conventional seed potato system interventions, based on rapid multiplication of disease free starter material.

Three research questions are answered through this second diagnostic study:
1. What are the current seed potato management practices of smallholder producers in Kenya, Uganda and Ethiopia?
2. How does the current seed potato system in the three countries function?
3. What innovative intervention pathways can be identified for seed potato system improvement that have not been pursued in earnest in the past?

The results of the study are presented in Chapter 3 of the thesis. The study confirms the common knowledge of potato specialists in the region that both virus diseases and bacterial wilt are endemic and likely contributors to the low yields observed in the research work presented in Chapter 2. The seed system study concludes that the current influence of the specialised seed potato chain on seed potato quality is limited. The vast majority of planting material used by farmers is obtained from neighbours or re-cycled from the own harvest. The research results reveal that this strategy by potato producers does make economic sense in the absence of affordable high quality seed potatoes combined with limited market security for consumption potatoes.

Considering this reality a dual strategy of simultaneously intervening in the specialised seed potato system and the local seed potato system is proposed. Commercial investments by seed potato entrepreneurs are considered an essential element for specialised seed system improvement. For improving the local seed potato chain, it is suggested to focus on opportunities for seed potato quality management by consumption potato growers themselves.

1.2.3 Improving the quality self-supply seed potatoes through positive selection

In response to the identification of seed potato quality management by ware potato farmers as a strategy complementary to specialised seed system building, action research was initiated in Kenya to assess the value of positive selection for East African potato systems. Positive selection is an old technology that was primarily used in the first stages of formal seed potato multiplication systems (Struik and Wiersema, 1999), before improved disease diagnostics and rapid multiplication through tissue culture became standard practice for starter seed production.

The action research effort on positive selection aimed to answer two research questions:
1. Can positive selection improve the quality of seed stocks commonly used by smallholder potato producers in Kenya?
2. Can positive selection be applied effectively by smallholder potato producers in Kenya?

Especially the first research question, but also part of the second question relate to the technical functioning of positive selection to improve the quality of consumer potato seed stocks. In Chapter 4 the more technical aspects of the three research questions are discussed. The chapter presents the data derived from a large number of farmer managed demonstration trials and shows that positive selection can contribute to increasing smallholder potato yields.

1.2.4 Investigating the mechanisms behind the effects of positive selection on seed potato quality

In response to the successful application of positive selection in farmer managed demonstration fields, questions were rising with regard to the mechanisms behind the effect of the technology. To investigate the supposition that the observed effect could

largely be contributed to a reduction in virus infection rates replicated trials were initiated around Kenya under widely different circumstances, and with different sources of seed potatoes. This component of the research programme aimed to answer the following five research questions:
1. Does positive selection reduce the virus load of seed potato stocks compared to the common farmer practice of selecting seed potatoes from the bulk of the harvest?
2. Can a reduction in virus load explain the yield increases observed compared to common farmer practice?
3. Is positive selection effective for different popular potato varieties?
4. Is there an interaction between the effect of positive selection and soil fertility?
5. Does the effect of positive selection depend on the source and quality of the seed potatoes with which a field has been planted?

1.2.5 Analysing the role of agricultural research in innovation

In retrospect the entire research effort, and specifically the research on the technology of positive selection, can be considered as a successful contribution of agricultural research to innovation. As a result of the research effort an alternative parallel intervention strategy has opened up for improving the potato sector in Sub-Sahara Africa. This research programme raises questions about the process of agricultural innovation and the role of agricultural research within this process. The research on positive selection provides a case to investigate the role of agricultural research in agricultural innovation as a participant observer. The case is of particular interest because the entry-point was an old, for some even obsolete, technology and the role of research went much beyond the validation or falsification of theoretical principles. A study on the role of research in the case of positive selection and the merit of the technology for potato system improvement is presented in Chapter 6. In the chapter three research questions are raised and discussed:
1. How did agricultural research contribute to the success of positive selection?
2. What were the internal conditions in the research organisation and the external factors that made this role effective?
3. What can be learned from this very specific case about the effective contribution of agricultural research to innovation in general?

1.2.6 Implication for seed potato system improvement and conclusions on the role of research

Chapter 7 will summarise and explore the implications of the results presented in the separate research chapters. First the chapter will explore the consequences for potato sector improvement in Sub-Sahara Africa. The findings presented in earlier chapters change the thinking about seed potato system improvement. The results hold important consequences for the design of interventions aimed at improvement of the quality of seed potatoes used in Sub-Sahara Africa. Suggestions are provided for applied and more fundamental research to further explore the new directions of thinking and opportunities for action opened up as a result of the research findings.

Chapter 7 will conclude by further exploring the insights that can be derived from the entire research programme with regard to the role of agricultural research in agricultural innovation. The experiences from this organically evolved research program are put in the larger perspective of applied agricultural research in developing countries.

References

Cromme, N., A.B. Prakash, N. Lutaladio, and F.O. Ezeta, (eds.) 2010. Strengthening potato value chains; Technical and policy options for developing countries, pp. 1-147. FAO, Rome, CFC, Amsterdam, Rome.

Ewing, E. 1981. Heat stress and the tuberization stimulus. American journal of potato research 58:31-49.

FAO. 2011. The state of food insecurity in the world; How does international price volatility affect domestic economies and food security? Food and Agricultural Organization of the United Nations, Rome.

IFAD. 2010. Rural Poverty Report 2011: New realities, new challenges: new opportunities for tomorrow's generation. International Fund for Agricultural Development, Rome.

IMF. 2011. Sustainaing The Expansion, Washington.

Scott, G.J., M.W. Rosegrant, and C. Ringler. 2000. Global projections for root and tuber crops to the year 2020. Food Policy 25:561-597.

UN-ESA. 2011. World Population Prospects: The 2010 Revision. [Online] http://esa.un.org/unpd/wpp/Analytical-Figures/htm/fig_11.htm (verified 21-02-2012).

UNCTAD. 2011. The least developed countries report 2011; The potential role of south-South cooperation for inclusive and sustainable development:194.

UNECA. 2010. Economic Report on Africa 2010: Promoting high-level sustainable growth to reduce unemployment in Africa. United Nations Economic Committee for Africa, Addis Abeba

UNECA. 2011. Economic Report on Africa 2011; Governing development in Africa - the role of the state in economic transformation, Addis Abeba.

UNFPA. 2011. State of the world population 2011; People and posibilities in a world of 7 billion. United Nations Population Fund, New York.

Woolthuis, R.K., M. Lankhuizen, and V. Gilsing. 2005. A system failure framework for innovation policy design. Technovation 25:609-619.

Worldbank. 2007. World Development Report 2008; Agriculture for Development The World Bank, Washington.

2 Improving potato production in Kenya, Uganda and Ethiopia: a system diagnosis

Peter Gildemacher[a,f], Wachira Kaguongo[a], Oscar Ortiz[a], Agajie Tesfaye[b], Gebremedhin Woldegiorgis[b], William Wagoire[c], Rogers Kakuhenzire[c], Peter Kinyae[d], Moses Nyongesa[d], Paul Struik,[e] Cees Leeuwis[e]

[a] International Potato Center, Nairobi, Kenya
[b] Ethiopian Institute of Agricultural Research, Holetta, Ethiopia
[c] Kachwekano Zonal Agricultural Research Institute, Kabale, Uganda
[d] Kenya Agricultural Research Institute, Tigoni, Kenya
[e] Wageningen University and Research Centre, Wageningen, The Netherlands
[f] Royal Tropical Institute, Amsterdam, The Netherlands

Published in: Potato Research (2009) 52:173-205

Abstract

Increased productivity of potatoes can improve the livelihood of smallholder potato farmers in Kenya, Uganda and Ethiopia and is required to meet the growing demand. This paper investigates the opportunities for potato system improvement that could result in improved productivity. Through a diagnosis of the potato systems in the three countries on the basis of surveys and stakeholder workshops, seed potato quality management, bacterial wilt control, late blight control and soil fertility management were identified as key technical intervention topics. For effective problem solving in these areas the functioning of the potato innovation system requires improvement to better deliver the functions of potato marketing as well as knowledge development and information exchange. Using a 'system failure framework' the shortcomings of the potato innovation system are identified and discussed and options for improvement are suggested.

2.1 Introduction

Potatoes are a source of both food and cash income in the densely populated highlands of Sub Sahara Africa. Through this double purpose the potato crop plays an important role in the rural livelihood system. Because of high prospects for growth of the market for fresh potatoes (Scott et al. 2000) the commodity could be a good starting point for rural development in Sub Sahara Africa, particularly under current conditions of increased cereal prices in the international markets.

Fuglie (2007) suggests broad fields of potato research and development that could be prioritized in different regions of the developing world. For effective targeting of research and development efforts a more detailed country or region specific analysis of the potato system and its potential opportunities and possible constraints is required. This analysis should not only identify important technological research areas, but also identify the weaknesses within the innovation system that have deterred and could continue to deter innovation of the potato sector taking place.

Kenya, Uganda and Ethiopia are among the ten African countries with the largest area cropped to potato (FAOSTAT 2006). These countries however differ substantially in terms of the integration of potato farmers in the market as well as the structure and functioning of the potato related innovation system. Together the three countries are representative of a large portion of the potato sector in Sub-Sahara Africa.

In this paper the current potato production systems in Kenya, Uganda and Ethiopia are diagnosed. Through a combination of stakeholder workshops (methodology adapted from Engel (1997) and Biggs and Matsaert (1999)) and two quantitative surveys among potato farmers, the potato production systems of Kenya, Uganda and Ethiopia were characterized and technical and innovation system related constraints that hinder potato productivity were identified. Using a system failure framework developed by Woolthuis et al. (2005), the strength and weaknesses in the potato related innovation system in the three countries are revealed. Finally, based on these findings, opportunities for potato system improvement in Kenya, Uganda and Ethiopia are discussed.

2.2 Methodology

Two surveys were conducted. The first survey examined the potato production practices and technologies of smallholder potato producers, while the second survey was conducted to assess the potato related knowledge and information system in the sub-region. Questionnaires were pre-tested and adapted before full implementation of the surveys by purposefully recruited and trained enumerators. Data from the surveys were coded and entered by each country team and centrally cleaned for all three countries and analysis using SPSS software. Costs of labour were calculated based on farmer estimates, and opportunity costs for labour were based on the average hired labour cost estimates. Economic analysis calculations were based on average yield figures over all cultivars and seasons, weighed for plot area.

2.2.1 Potato practices and technology survey

The potato practices and technology survey focused on documenting management practices and technology used by potato producers in Kenya, Uganda and Ethiopia. This survey yielded the information on productivity and economics of potato presented in this paper.

In Kenya, data collection took place between 10 and 29 October 2005. After a rapid appraisal of the potato system in Kenya, Meru Central and Nyandarua districts were selected as sample districts as they were considered to best represent the whole of the Kenyan potato production system. A district is an administrative topographical unit which is further subdivided into divisions, locations and sub-locations, which is the smallest administrative unit. Six farmers were randomly chosen in half of the sub-locations within each location in the sampled districts to get a satisfactory number of sample farmers. The sample households were randomly picked from a list of all farm households in the village, provided by a village elder. In total 251 farmers were successfully interviewed, 100 in Meru Central district and 151 in Nyandarua district.

In Uganda the survey was implemented between 1 and 25 November 2005. Kabale and Kisoro districts were selected to represent the potato farming system in Uganda. All the 4 counties and the 25 potato producing sub-counties in Kabale and Kisoro districts were included in the study. One parish was randomly selected from each sub-county and one village randomly selected within each parish. Six households were picked at random in each sampled village from a list of households provided by a village elder to assure a sufficiently large and representative sample. In all, 144 farmers out of 150 randomly selected were successfully interviewed. In addition 89 farmers were randomly picked from selected farmer groups who had participated in earlier potato related research and development activities.

In Ethiopia data collection took place between 5 February and 27 March, 2007. Three major potato producing districts (woredas) were selected, Jeldu in West Shewa zone, Degem in North Shewa zone and Banja Shikudadin in Awi zone, as a cross section of potato production in the country. Within these districts six households were randomly selected within each kebele (village), resulting in 220 households that were successfully surveyed. In addition, 116 households of farmers that had participated in activities by the Ethiopian potato development project partner were selected from all participants. In Ethiopia therefore, 336 households were surveyed.

Data from the selectively sampled interviewees who had been participants in research and development activities in Uganda and Ethiopia were only included in calculating total crop coverage and marketing figures which were considered independent of participation.

2.2.2 Knowledge and information survey

Second, a survey was conducted among potato producers in Kenya, Uganda and Ethiopia to assess the relative importance of different sources for information on potato farming practices and marketing.

In Kenya, Bomet and Nyandarua were selected as sample districts. Nyandarua represents potato farming for the wholesale ware potato market, whereas Bomet is important for the production of potatoes for the crisp processing. In Bomet district, farm households were randomly selected equally among its six divisions. Interviews were conducted in May 2004. In Nyandarua farmers were selected randomly in Kipipiri and North Kinangop divisions, which are considered to be representative of potato farming in Nyandarua district. Within the divisions, locations, sub-locations and villages were randomly, selected. Within the villages households were selected through a random transect walk, selecting each fifth household. Interviews were conducted in October 2004.

In Uganda, Kabale district, which produces the highest bulk of potatoes (Low 1997) was selected as sample district. Ten out of 19 sub-counties in this district were selected randomly for the survey. Within each parish in the selected sub-counties a village was sampled. Farm households were selected randomly from a list of households provided by a village elder. The survey was implemented in September 2005.

In Ethiopia, Jeldu, Dendi and Wolemera districts were selected in the West Shewa zone, Degem district in the North Shewa zone and Alemaya district in the East Shewa. These districts were chosen because of the importance of potato in the farming system, their differences in potato farming practices, and because they are intervention areas for potato agro-enterprise development by various development organizations. Within these districts peasant associations (the lowest administrative unit) were randomly selected and within this farm households were picked at random from a list of all farmer families provided by the local office of the Ministry of Agriculture. The interviews were conducted in June and July, 2004.

In Kenya, Uganda and Ethiopia 97, 211 and 646 farm households were interviewed, respectively. The total target of interviewed farmers was adapted to the resources available in each country. Farmers were interviewed using a questionnaire that was developed by the researchers from the 3 countries, followed by appropriate adaptation for each country.

2.2.3 Stakeholder meetings to assess potato related innovation system

Stakeholder workshops were organized to identify constraints and opportunities in the potato sector in Kenya, Uganda and Ethiopia. An assessment of the system was made from an innovation system perspective, focussing primarily on the interrelations between the stakeholders and their respective roles in knowledge development and information exchange. In Kenya, two single day stakeholder workshops were conducted in both Bomet (11 June 2004 and 13 January 2005) and Nyandarua (19 October 2004 and 16 November 2005) districts. In Uganda, a single day workshop with potato stakeholders from Kabale district was organized on 9 March 2005. In Ethiopia, the meeting consisted of a single, event from 21-23 July 2004, with representatives of potato sector stakeholders from the Alemaya, Galessa, Jeldu and Degem districts. Stakeholder categories present in the meeting varied per country as a result of different responses to invitations to attend.

Workshop participants were grouped together according to stakeholder category, for example ware potato farmers, seed potato farmers, public extension staff, representatives of non-governmental organizations intervening in rural development (NGOs), processors, transporters or agro-input suppliers. Stakeholder categories present in the meeting varied per country as a result of different responses to invitations to attend. All groups analysed their own role and the role of other stakeholders in the potato chain and an Actor Linkage Matrix of all interactions was constructed, adapted from a method described by Biggs and Matsaert (1999). First every stakeholder group identified its interactions with other stakeholder groups in the potato chain. Consecutively the constraints in these interactions were identified. The actor linkage matrix was constructed by the workshop facilitators (Figure 2.1) and the opinions of the different stakeholder groups about each other were presented in a plenary for discussion.

In Kenya, in the second workshop in both Bomet and Nyandarua the problems identified in the first workshop were prioritized through a ranking exercise. Next, solutions to the most important constraints were discussed in mixed groups consisting of different stakeholder categories. The group results were reported back to the plenary for further elaboration.

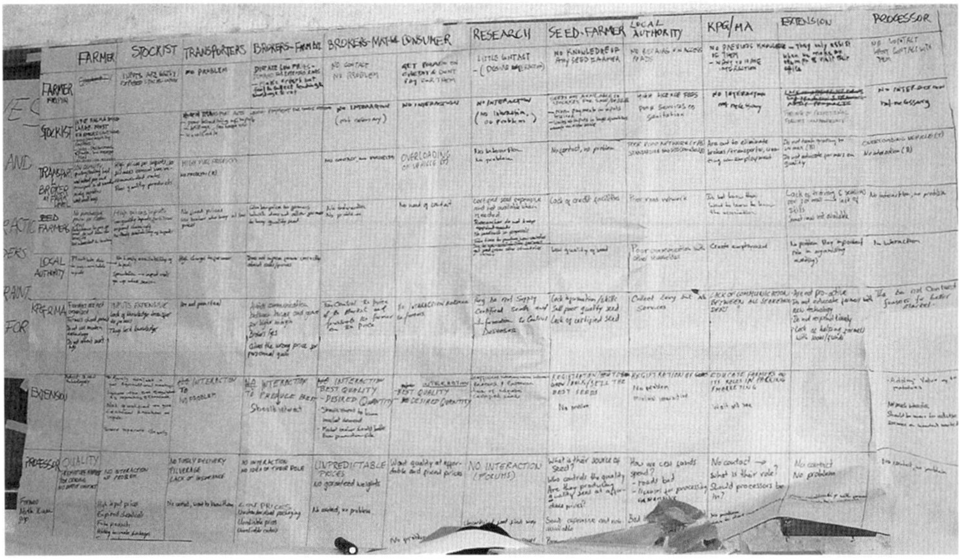

Figure 2.1: Actor Linkage Matrix in Nyandarua, Kenya, 2004.

2.3 Potato production system characterization

2.3.1 Trends in potato production and productivity

Potato production is projected to grow by 2.7% a year globally until 2020 (Scott et al. 2000), a growth that exceeds all other major food crops. For Sub Saharan Africa the same authors project an annual growth of demand of 3.1%. 2 shows the growth of the estimated area under potato in Kenya, Uganda and Ethiopia from 1996 until 2006. A steady increase in area over time can be observed: the average increase is 4.3% per year for Kenya and 7.0% per year for Uganda. For Ethiopia, FAO reports a modest growth in area of 2.6% per year, which contrasts sharply with data gathered by the potato research station of the Ethiopian Institute of Agricultural Research for the years 1995–2000 (EIAR, unpublished data) which shows an explosive growth in area of 22% per year (Figure 2.2).

For the same period during which the area under production increased rapidly, limited change in productivity per unit area can be observed (Figure 2.3). This clearly shows that the growing demand is met by area increase, rather than by yield increase.

Potato productivity estimates in the surveys indicate yields in Kenya of 9.1 Mg ha^{-1} (Table 2.1) whereas in Uganda farmers estimated average yields of 5.8 Mg ha^{-1} which is well below the yields estimates provided by the FAO (Figure 2.3). In Ethiopia farmers estimated to yield an average of 7.9 Mg ha^{-1}, which is in line with the FAO data.

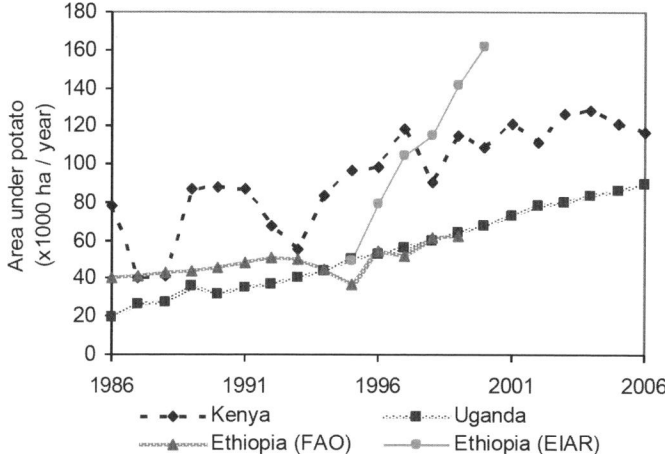

Figure 2.2: Area under potato in Kenya, Uganda and Ethiopia from 1986–2006
Source: Faostat, 10-2007; EIAR 2007

Table 2.1: Average potato productivity as estimated by farmers in Kenya, Uganda and Ethiopia (Mg ha-1).

	Kenya	Uganda	Ethiopia
Average production	9.1	5.8	7.9
Median	7.7	4.2	6.0
Standard Error of Mean	0.35	0.43	0.44
N	249	128	177

Source: potato practices and technology survey

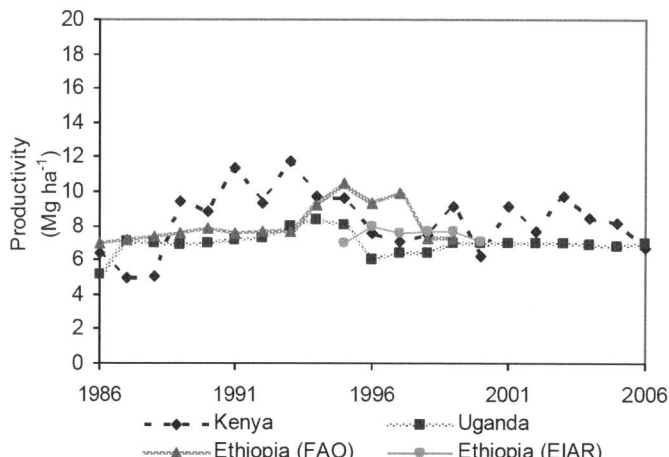

Figure 2.3.: Productivity of potato (Mg ha-1) in Kenya, Uganda and Ethiopia from 1986–2006
Source: Faostat, 10-2007; EIAR 2007

2.3.2 Potato production seasons

Most potato farming occurs under rainfed conditions. Consequently the major cropping seasons follow the rainy seasons (see Figure 2.4). In Kenya two seasons with ample rainfall for potato cultivation could be identified (see Figure 2.4). The length of the seasons depended on the region. Limited off season production occured in areas higher than 2,000 m above sea level, on slopes receiving intermittent rainfall and mist which combined with residual moisture could support a potato crop in most years. Irrigated out of season potato farming was virtually limited to Abo West division in Meru Central district, on the slopes of Mt. Kenya, where farmers use small sprinklers which are connected to upslope streams, and operate on force of gravity. Most potato farmers in Abo West timed potato production in a manner that they can harvest potatoes in between the supply peaks of the rainfed crop.

In Uganda, two main production seasons were identified, coinciding with the short and long rainy seasons. However, in Kabale district, a third cropping season can be recognized after the short rainy season (Figure 2.4). During this time potato is planted in valley bottoms or drained wetlands. The crop is supported by the residual moisture available in the rich organic soil or drainage water coming from the surrounding hills.

In Ethiopia in most of the potato farming zones two rainy seasons can be identified, the main (Meher) season and a short rainy season (Belg). Potato farming during the Belg season is done either at high altitude where evapotranspiration is low and rainfall higher than average in the country. However, in many areas the Belg season is short and unreliable and supplementary irrigation is imperative. Considerable differences in rainfall patterns occur between and even within potato growing zones in Ethiopia, which means there are many exceptions to main potato growing seasons presented in Figure 2.4.

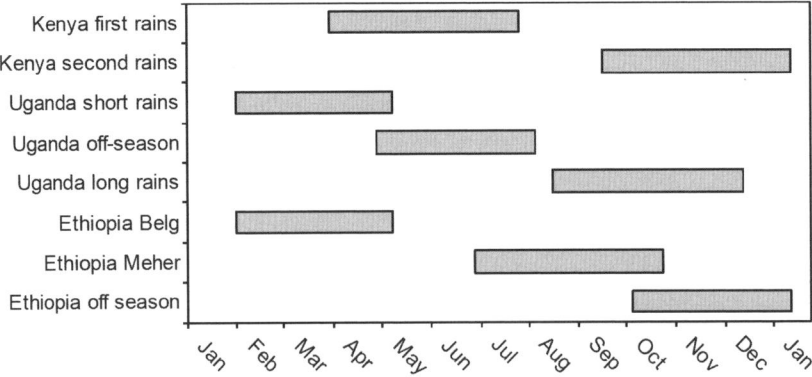

Figure 2.4: Main potato growing seasons in Kenya, Uganda and Ethiopia

2.3.3 Importance of the potato crop in the farming system

In Kenya, the majority of sampled farmers cultivated potato twice a year, during the main rainy seasons. Potato farmers dedicated more than a third of their arable land to potato in both districts during both seasons (Table 2.2. During the first rainy season of 2005 potato covered 30% of the total cropped area in the sample districts (Figure 2.5).

In Uganda, the majority of sampled farmers grew potato twice a year, dedicating 24-32% of their arable land to potato depending on the season and district (Table 2.2.). During the short rainy season of 2005, 24% and 23% of all arable land in Kabale and Kisoro respectively was cropped to potato (Figure 2.5).

In Ethiopia the main potato growing season depended on the zone (Table 2.2.). In West Shewa the main production season was during the Belg season, in North Shewa during the Meher season, and in Awi zone more farmers grow potato during the Belg and off-season than during the Meher season. Of all crops grown during the Meher season of 2006, potato occupied 7, 8, and 13% of all cropped land in North Shewa, Awi and West Shewa zones, respectively (Figure 2.5).

Table 2.2: Percentage of farms cultivating potato, average size of potato fields and the proportion of arable land devoted to potato on farms growing potato during different seasons in Meru Central and Nyandarua districts of Kenya (2004-2005), Kabale and Kisoro districts of Uganda (2004-2005) and the West Shewa, North Shewa and Awi zones of Ethiopia (2005–2006).

	Farms growing potatoes (%)	Potato fields (ha/farm)	Fraction potatoes (% of farm)
Kenya			
Meru central (n=100)			
Season 1 (April-July)	77	0.32	42
Season 2 (Oct-Jan)	95	0.34	38
Off season (Jul-Oct)	9	nd	nd
Nyandarua (n=151)			
Season 1 (April-July)	93	0.40	35
Season 2 (Oct-Jan)	99	0.47	37
Off season (Jul-Oct)	0		
Uganda			
Kabale (n=169)			
Season 1 (April-July)	91	0.27	25
Season 2 (Oct-Jan)	72	0.28	26
Off season (Jul-Oct)	28	nd	nd
Kisoro (n=61)			
Season 1 (April-July)	80	0.23	28
Season 2 (Oct-Jan)	92	0.25	32
Off season (Jul-Oct)	8	nd	nd
Ethiopia			
West Shewa (n=138)			
Belg season (Feb-May)	83	0.27	20
Meher season (Jun-Oct)	53	0.43	34
Off season (Oct-Jan)	7	0.17	13
North Shewa (n=104)			
Belg season (Feb-May)	9	0.14	8
Meher season (Jun-Oct)	86	0.18	9
Off season (Oct-Jan)	17	0.16	6
Awi (n=94)			
Belg season (Feb-May)	84	0.24	29
Meher season (Jun-Oct)	40	0.21	21
Off season (Oct-Jan)	65	0.20	23

Source: potato practices and technology survey
nd = no data

Figure 2.5: Total proportions of arable land cropped to different crops during the first rainy season in Kenya and Uganda in 2005 and the Meher season in Ethiopia in 2006

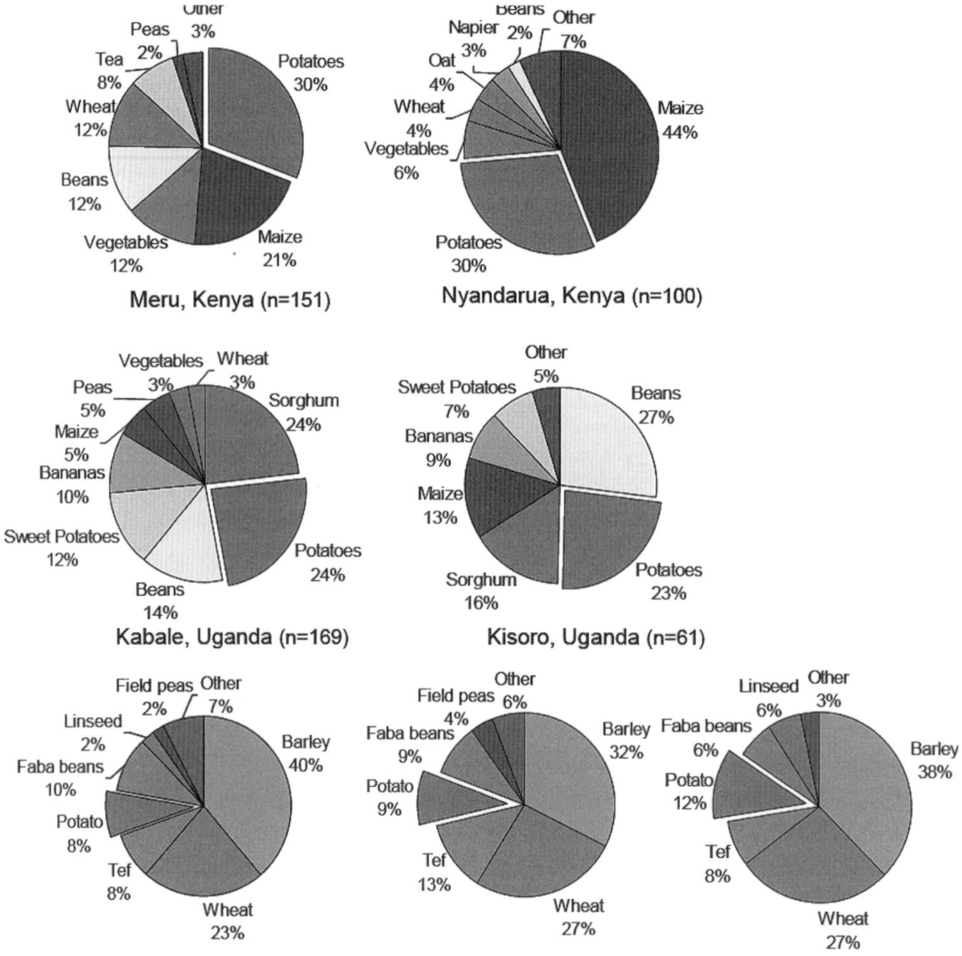

Meru, Kenya (n=151)
- Potatoes 30%
- Maize 21%
- Vegetables 12%
- Beans 12%
- Wheat 12%
- Tea 8%
- Peas 3%
- Other 2%

Nyandarua, Kenya (n=100)
- Maize 44%
- Potatoes 30%
- Other 7%
- Vegetables 6%
- Wheat 4%
- Oat 4%
- Napier 3%
- Beans 2%

Kabale, Uganda (n=169)
- Sorghum 24%
- Potatoes 24%
- Beans 14%
- Sweet Potatoes 12%
- Bananas 10%
- Peas 5%
- Maize 5%
- Vegetables 3%
- Wheat 3%
- Other 7%

Kisoro, Uganda (n=61)
- Beans 27%
- Potatoes 23%
- Sorghum 16%
- Maize 13%
- Bananas 9%
- Sweet Potatoes 7%
- Other 5%

(Ethiopia site 1)
- Barley 40%
- Wheat 23%
- Faba beans 10%
- Potato 8%
- Tef 8%
- Other 7%
- Field peas 2%
- Linseed 2%

(Ethiopia site 2)
- Barley 32%
- Wheat 27%
- Tef 13%
- Potato 9%
- Faba beans 9%
- Other 6%
- Field peas 4%

(Ethiopia site 3)
- Barley 38%
- Wheat 27%
- Potato 12%
- Tef 8%
- Faba beans 6%
- Linseed 6%
- Other 3%

Source: potato practices and technology survey

2.3.4 Use of agricultural inputs

Farm yard manure can supplement inorganic fertilizer to maintain soil fertility. It was however only widely used in Kenya, where 45% of the farmers indicated having used manure on their last potato crop, compared to 18 and 26% of potato farmers in Uganda and Ethiopia respectively (Table 2.3). The average amount of farm yard manure applied in Kenya by those farmers using it, contained 48 and 13 kg of N and P respectively, if using the farm yard manure composition figures presented by Lekasi et al. (2003).

Table 2.3 shows large differences amongst the countries in utilization of fertilizer in potato farming. In Kenya, the vast majority of farmers used mineral fertilizers in potato farming, and in substantial quantities, although the rates used were on average less than half of the recommended amounts of N and P of 90 and 230 kg per hectare in the form of DAP (18:46) (Source: KARI-Tigoni). In Uganda, fertilizer use on potatoes was virtually absent, which shows there was no change in practices since this was last assessed by Low (1997). In Ethiopia more than half of the potato farmers indicated to have used fertilizers on their last potato crop, however in lower quantities than Kenyan farmers.

Table 2.3: Percentage of potato farmers in Kenya, Uganda and Ethiopia using farm yard manure and mineral fertilizer and the average amounts applied by these users.

	Farmyard manure (FYM)			Fertilizer				
	Farmers using FYM (%)	FYM applied (kg/ha)	s.e.	Farmers using fertilizer (%)	N applied (kg/ha)	s.e.	P applied (kg/ha)	s.e.
Kenya (n=251)	45.0	4327	512	87.8	43.3	2.0	101.4	4.7
Uganda (n=141)	17.7	2207	606	4.7	37.6	18.9	46.9	45.1
Ethiopia (n=287)	26.1	3006	317	57.2	30.6	2.5	33.4	2.3

Outliers in farmer estimates of applied amounts were removed by skimming the top 5% estimates.
Source: potato practices and technology survey

2.3.5 Seed potato source and renewal

Low seed potato quality is a major constraint of potato farmers in Kenya, Uganda and Ethiopia (Gildemacher et al. 2009b). The seed potato system in the three countries is characterized by a limited availability and use of commercially traded high quality seed potato.

Assessment of seed potato renewal revealed that in Kenya 41%, in Uganda 26% and in Ethiopia 44% of the farmers periodically renewed their seed stock. On average those farmers who did renew their seed did so after 6 seasons in Kenya, 7 seasons in Uganda and three seasons in Ethiopia. The most important source of seed potato in Kenya were neighbours, while in Uganda and Ethiopia the village market was the dominant source. Specialized seed growers were a source of seed in only 2% of the cases in Kenya, compared to 10% in Uganda and 16% Ethiopia (Table 2.4).

Table 2.4: Seed potato renewal frequencies and sources of seed in Kenya, Uganda and Ethiopia.

	Kenya	Uganda	Ethiopia
Seed renewal frequency			
Farmers renewing seed stock (%)	41	26	44
Average renewal period	6	7	3
N	250	141	186
Source of seed last renewal (%)			
Neighbour	94	34	14
Market	5	56	69
Specialized seed grower	2	10	16
N	311	157	301

Source: potato practices and technology survey

2.4 Potato marketing system characterization

2.4.1 Home consumption versus marketing

The study also confirmed that potato is a dual purpose crop in the three countries studied. It serves both the household staple food requirement and as a source of cash income. On average, potato farmers in Kenya produced more potato tubers than their counterparts in Uganda and Ethiopia. Farmers in Kenya marketed over 80% of their harvest after satisfying their home consumption and seed potato needs, while in Uganda 71% of the produce was sold compared to 61% in Ethiopia (Table 2.5).

Table 2.5.: Marketed potato yield versus home consumption in Kenya, Uganda and Ethiopia.

	Kenya (n=96)		Uganda (n=154)		Ethiopia (n=419)	
	Weight (kg) per household per season	%	Weight (kg) per household per season	%	Weight (kg) per household per season	%
Ware sold	2,899	77	753	61	528	48
Seed sold	165	4	112	9	146	13
Ware home	327	9	191	16	300	27
Seed home	352	9	170	14	126	11
Total	**3,743**		**1,226**		**1,107**	
Total market	3,065	82	865	71	673	61
Total home	679	18	361	29	434	39

Source: potato practices and technology survey

Kenyan farmers sold a smaller proportion as seed potato, but tended to keep more seed potatoes for their own production than the farmers in Uganda and Ethiopia. The survey shows that farmers in Kenya tended to plant bigger seed potatoes than their peers in Uganda and Ethiopia. Furthermore the combination of a higher productivity and a larger average area under the crop in Kenya compared to Uganda and Ethiopia resulted in a larger proportion of surplus potatoes for the market after satisfying the home needs for seed and consumption potatoes (Table 2.5).

2.4.2 Potato marketing channels

In Kenya, the bulk of potato (87%) was sold to traders direct from the field, mostly assisted by field brokers (Table 2.6). The field brokers scout for potato fields that are ready to be harvested and negotiate deals ahead of the arrival of a transporter. In Uganda roughly two thirds of the ware potatoes were traded directly from the field, while 25% of the ware potato produce was sold through village markets. In Ethiopia most ware potato farmers took their produce piecemeal to village markets, often on horseback or hand carried, where it was sold to wholesalers.

The data indicated that Kenya and Uganda had a better developed marketing system, where farmers can sell directly from their fields to brokers. This reduces the need for on-farm ware potato storage, minimizes the efforts farmers have to make to sell their produce and maximizes the sum received at once thus facilitating meaningful investment of their revenue. However, the potato farmers in the stakeholder workshops in Kenya indicated that this marketing system is far from being perfect. It gives field-level brokers a leverage to offer low farm-gate prices to increase their own profit margin at the cost of farmer efforts. However, the village brokers play a significant role as a link between the urban-based potato traders and the potato farmers.

In the three countries studied, farmers traded their produce predominantly as individuals. Only in Ethiopia very few farmers sold their potato produce through a farmer cooperative or NGO (Table 2.6).

Table 2.6: Marketing outlets used by potato producers in Kenya, Uganda and Ethiopia.

	Potato marketing channel (%)		
	Farmgate to trader / broker	Village market	Other[a]
Kenya (n=105)	87	8	6
Uganda (n=175)	68	25	8
Ethiopia (n=435)	4	88	8

Source: knowledge and information survey
[a] Includes roadside sale, farm gate retailing, farmer cooperative and NGO

2.5 Potato production economics

2.5.1 Profitability of potato production

In Meru Central (Kenya), farmers invested more in their potato production than in other sampled areas (Table 2.7). A relatively high average yield and fairly high prices resulted in a positive net margin, and a high return on family labour. In Nyandarua yields were comparable to those in Meru Central, but the prices farmers received for their produce were very low, and in spite of lower cash investments their net margin was negative. The low investments in fertilizer, fungicides and hired labour in Nyandarua compared to Meru, and the use of more family labour are an understandable reaction to the low prices. In this manner potato farmers do get a positive return on their cash investment, providing them with some cash income as well as food from their potato production. Return on their labour however, is lower than the opportunity cost.

Farmers in both Kabale and Kisoro districts in Uganda obtained net margins comparable to Meru Central in Kenya (Table 2.7) although mean yields were lower in Uganda than in Kenya. In Uganda, cash investments in potato production were lower than in Kenya as farmers hardly used fertilizer, labour was cheaper in Uganda and farmers used more family labour than in Kenya.

In Ethiopia, specifically farmers in North Shewa were making good profits compared to all other sampled districts. This is the combined result of much higher yields compared to West Shewa and Awi zone, low investments in inputs and a relatively good price. In West Shewa both yields and prices were lower, while in Awi zone prices were good, but yields very low.

Table 2.7: Average production costs and revenues of potato production in major potato production areas of Kenya, Uganda and Ethiopia, 2005-2006[a].

	Kenya (n=251)		Uganda (n=144)		Ethiopia (n=220)		
	Meru Central	Nyandarua	Kabale	Kisoro	West Shewa	North Shewa	Awi
Cash investments							
Cost fertilizer (US$/ha)	118.40	79.49	1.81	7.27	21.12	4.12	5.84
Cost fungicides (US$/ha)	33.21	16.22	25.31	26.34	38.57	2.85	0.00
Hired labour used (US$/ha)	104.34	103.32	45.61	50.82	120.38	110.39	85.51
In kind investments							
Cost manure (US$/ha)	30.67	5.86	0.83	1.68	3.23	2.42	1.99
Cost seed (US$/ha)[b]	123.37	72.94	96.56	72.37	65.84	88.72	144.10
Family labour investments							
Family labour used (days/ha)	97.83	210.57	227.81	211.59	117.00	125.99	149.16
Family labour (opp. costs; US$/ha)[c]	100.38	197.04	107.87	85.63	116.46	149.87	156.27
Revenues							
Yield (Mg ha-1)	8.83	9.21	5.25	6.83	7.69	12.33	5.14
Price (US$/t)	80.16	45.10	88.09	72.06	62.44	79.12	91.22
Gross margin (US$/ha)	707.85	415.34	462.45	492.14	480.09	975.73	468.89
Net margin (opp. cost; US$/ha)	**197.48**	**-59.52**	**184.45**	**248.04**	**114.50**	**617.36**	**75.19**
Net margin (no opp. Cost; US$/ha)[d]	297.86	137.51	292.33	333.67	230.96	767.23	231.46
Return on family labour (US$/day)[e]	4.13	0.97	2.08	2.99	1.97	6.09	1.55
Return on cash investment (US$/US$)[f]	1.77	1.09	5.36	4.83	1.67	7.31	4.13

[a] Figures presented are calculated from the average of all valid farmer estimates. Farmers have estimated costs and revenues per plot per season, which was then weighed for area over seasons, varieties and plots.
[b] Cost of seed is put at the average price of ware potato because of widespread use of farm-saved seed.
[c] Opportunity cost of labour is put equal to average estimated cost of hired labour
[d] No opportunity costs are calculated for family labour
[e] Net margin divided by days of family labour
[f] (Gross margin minus cash investments) / cash investments; only fertilizer, fungicides and hired labour are considered cash investments

Source: potato practices and technology survey

2.5.2 Return on cash investment

The return on cash investment was more than 100% in all sampled districts (Table 2.7). It can be concluded that the smallholder potato farmers are risk averse in terms of their cash investment. An economic strategy can be observed in which growers minimize the risk of cash losses as a result of their potato production. The producers are reluctant to invest their scarce cash resources in seed potatoes, fertilizer and fungicides, even though this could increase their profits and, at least partially, rely on their own labour and seed potatoes, and minimize the use of fertilizer in their production system. The risk-adverse decision making with regard to cash investments is an economically sensible response to insecurities regarding potato diseases, drought and potato prices.

2.5.3 Return to family labour

In all sampled areas, except Nyandarua district in Kenya, the return on family labour in terms of cash income was higher than the opportunity cost for their labour. This confirms that potato production can effectively contribute to the cash requirement of smallholder households and provides profitable employment to smallholder farmers.

The negative net margin in Nyandarua does not mean that farmers are making irrational economic decisions by farming potatoes. The net margin was derived after giving a value to their family labour at the average rate of hired farm labour in their district. Furthermore a cost was attached to their seed potato. In the first place the rural economy is far from perfect, and a productive alternative use of the family labour may not be possible. Secondly farmers use their own recycled seed potatoes as seed, which have no cost to the farmer, and are not easily marketable. Thirdly farmers in Nyandarua did have a positive return on their cash investment thus providing the household with scarce cash resources. Lastly their potato crop provides the household with food which would otherwise have required cash to purchase.

2.6 Identification of technical yield reducing factors and innovation priorities

2.6.1 Improving productivity

Over the last 20 years potato production increase in East Africa has been driven by area expansion as previously observed. However further potato production increase through area expansion will become increasingly difficult. Further area expansion in the highlands could fuel encroachment on the limited remaining highland forests in Eastern Africa. A high proportion of potato production within the rotation is not sustainable from the point of view of soil-borne pest and disease management and soil structure and fertility management (Struik and Wiersema 1999). Especially in Kenya and Uganda the proportion of land cropped to potato is already high (Figure 2.5). In the North Shewa zone in Ethiopia, there appears to be more opportunity for additional potato production within the rotation (Table 2.2 and Figure 2.5).

The most feasible manner in which the growing demand for potatoes can be satisfied is through increased productivity. Occasionally, substantially higher yields per unit area than presented in Figure 2.3 and Table 2.1 were obtained by farmers under the same rainfed production circumstances as neighbouring farmers. Gildemacher (2007a) report 12 t ha^{-1} in Kenya in farmer managed trials using normal farmer practices. Average yields from a large number of trials in the region were reported at 20 Mg ha^{-1} (Gildemacher et al. 2007b). The average yields of the top 10% in the potato practices and technology survey are 17.4, 12.0 and 17.0 tonnes of potatoes per hectare in Kenya, Uganda and Ethiopia respectively, which is twice as much as the national averages presented in Table 2.1.

This means that there is a potential to increase potato yields substantially compared to the current yield levels, especially considering the profitability of potato production demonstrated in Table 2.7. Based on the outcomes of the stakeholder workshops, combined with the quantitative data from the two surveys, four key technical areas that need to be addressed to increase productivity have been identified by the authors:
1. Seed potato quality management,
2. Management of bacterial wilt (caused by *Ralstonia solanacearum*),

3. Management of late blight (caused by *Phytophthora infestans*),
4. Soil fertility management.

These areas will be discussed below.

2.6.2 Seed potato quality management

One of the main technical issues raised by farmers during the stakeholder workshops was the limited availability and use of quality seed potatoes (Gildemacher et al., 2006a). The smallholder potato producers were giving both scientists and extension staff the challenge of assuring the supply of high quality seed potatoes. Ware potato farmers, seed multipliers and scientists agreed that improving the quality of seed potatoes used by the smallholder potato producers was an important requirement for the increase of productivity.

Considering the short crop rotations that farmers practice, seed potato quality is an important factor in improving the sustainability of production. The management of soil-borne pests and diseases depends on the combination of ample crop rotation and the use of quality disease free seed potatoes.

Without periodic renewal of the seed potato stock, tuber borne diseases build up in the seed stock that farmers use. Especially virus diseases are suspected to cause widespread and large yield reductions. In an extensive training program on virus diseases researchers in Kenya (Gildemacher et al. 2007a) have learned that the majority of ware potato producers did not recognize virus infection as a disease. The majority of farmers perceived virus symptoms a normal characteristic of a potato crop. Virus diseases were not mentioned by farmers as one of the important diseases that hamper potato production, but farmers in the stakeholder workshops indicated they lack a reliable source of timely available and affordable high quality seed potatoes. This indicates that farmers to some extent are aware of the importance of high quality seed potato to obtain high yield. The prevalence of potato viruses became very clear as a result of a survey sampling seed potatoes sold at rural markets in Kenya. Only 0.4% of all sampled tubers was free of the main four potato viruses (Gildemacher et al, 2007b).

The major challenge is that in the current potato system seed potatoes are not routinely bought by farmers as an input into their farming system. Shifting towards a mechanism where potato producers routinely renew their seeds to keep the yield potential high means a major shift in the potato system in the three countries. Before offering technological or system innovations to improve the quality of seed potatoes used by smallholder farmers a more thorough study of the functioning of the current seed potato system is required, which is reported separately (Gildemacher et al., 2009b).

2.6.3 Bacterial wilt control

Bacterial wilt (*Ralstonia solanacearum*) forms another major threat to intensive potato production in the East African highlands. Surveys in Kenya showed that 59% of the sampled potato plots in the upper highlands and up to 86% of potato fields below 2000 m above sea level were infested with bacterial wilt (Wakahiu et al. 2007). Farmers in both Kenya and Uganda mentioned bacterial wilt most frequently when asked about pest and disease problems in potato production during the survey (Table 2.8). Earlier Turkensteen (1987) identified bacterial wilt and late blight as the two most important pathogens in African potato growing systems. In the stakeholder workshops in Kenya and Uganda bacterial wilt and its management was a hotly debated topic.

The bacterial wilt problems are expected to increase as a result of shortening rotation. The disease survives in the soil for several seasons, and one essential component of bacterial wilt management is denying the bacteria a host by not growing potatoes or any other host crop for several seasons, combined with a strict removal of volunteer potato plants (Lemaga et al. 2005).

The management of bacterial wilt is further complicated by the lack of reliable seed sources and the heavy reliance on farm saved seed potatoes as planting material, which results in frequent re-infection of fields. The bacteria survive in non-symptomatic tubers that are stored for future planting, and the disease spreads once these tubers are planted. Bacterial wilt control as a point of intervention to improve potato productivity will have to include a seed potato quality management component, and is thus strongly related to the topic above.

Table 2.8: Potato disease problems prioritized by farmers in Kenya and Uganda, 2004-2005

	Kenya(n=99)	Uganda (n=155)
Late blight (*Phytophtora infestans*)	49	119
Bacterial wilt (*Ralstonia solanacearum*)	71	132
Other	1	2

Source: knowledge and information survey

2.6.4 Late blight control

Late blight (*Phytophthora infestans*) is a major potato disease that can result in the total destruction of the crop. Average yield losses were 20% over a large number of on-station trials in six countries in Eastern and Central Africa on potato cultivars with levels of host resistance to the pathogen (Gildemacher et al. 2007b). In both Kenya and Uganda, late blight was mentioned most frequently after bacterial wilt as the major disease constraint of potato (Table 2.8). This re-confirms the findings by Nyankanga et al. (2004), Turkensteen (1987) and Low (1997).

Late blight is the major cause of use of pesticides on potatoes in Kenya (Nyankanga et al. 2004), Uganda (Low 1997) and Ethiopia. Even if late blight is kept under control by farmers through the use of fungicides, there are both economic and ecological consequences of the pathogen. These may not be considered by farmers when ranking potato diseases. Although late blight is considered second in importance to bacterial wilt by farmers in the survey (Table 2.8), it may well present an even more important yield and revenue reducing factor. This is confirmed by its ranking as the priority crop management need by potato scientists from the African (Fuglie 2007).

2.6.5 Soil fertility management

There were large differences in the soil fertility management practices among the three countries. In Kenya, substantial amounts of nutrients were added to the soil in the form of farm yard manure and mineral fertilizer, mainly di-ammonium phosphate (DAP 18:46:0), while in Ethiopia the use of mineral fertilizers was much less important. In Uganda very few farmers used any soil amendments, either organic or mineral, in potato production (Table 2.3).

Potato yields in Kenya were higher than in the other two countries (Table 2.1). This could well be the result of better soil fertility management than in Ethiopia and Uganda. The only exception was in North Shewa, Ethiopia, where limited quantities of fertilizers were used, but yields were the highest of all three countries. This may be related to a more extensive farming system, with more rotation opportunities and a longer fallow due to larger farm sizes.

Soil fertility management did not feature prominently as a technical constraint in the stakeholder workshops. Farmers considered high fertilizer prices as a constraint, especially in Kenya. In spite of this lack of concern shown by farmers regarding soil fertility management, potato yield levels indicate that they are far below actual potential and that attention is needed in this field in the three countries to increase productivity in a sustainable manner. This opinion is shared with the scientists involved in the priority setting done by Fuglie (2007), who rated this topic as the priority in Sub Saharan Africa.

A specific area of interest is the study of how to optimise the use of limited amounts of fertilizer. Even in Kenya the average amount of mineral fertilizer used by farmers is still less than half the rate recommended. Optimizing the manner in which the amounts of fertilizer that potato farmers in Sub Sahara Africa can afford are best combined with the usually also inadequate organic sources of nutrients available to them is an important area for further research (Vanlauwe and Giller 2006).

2.7 Potato innovation system

The four areas that were identified as technical intervention points for improving potato productivity will help in targeting research and development efforts for the potato sector in the region. Effective problem solving, however, requires more than problem identification. The success of research and development efforts depends largely on the context in which they are initiated. According to the stakeholders in the workshops, two non-technical elements need specific consideration, and these were the marketing system and the knowledge and information system. Both can be seen as 'functions' of the potato innovation system. In this section the functioning of the potato innovation system is analysed and the question is answered how the potato related innovation system can support improvement of the potato sector in general, and specifically assure an environment in which the above identified technical innovation priorities can be tackled.

2.7.1 Potato marketing

Improving productivity alone will not always result in a dramatic improvement of profitability (Table 2.7). With the exception of Meru Central in Kenya, most farmers have a production strategy that reduces risk through minimizing cash investment in the potato production thus fulfilling their dual objective of cash income and household food security. When considering the current profit margins, this may well be an optimal production strategy. Under the low price expectations in Nyandarua, a farmer would be reluctant to invest cash to improve his potato production. Prices in West Shewa in Ethiopia are also fairly low compared with the other sample regions, even though it is in the relative vicinity of Addis Ababa.

During the stakeholder workshops in the three countries topics related to potato marketing featured prominently (Gildemacher et al. 2009a), confirming that potato marketing is a serious constraint for farmers and a major driver in decision. Any attempt to increase potato productivity will require some form of investment by the potato farmers. It may require cash investments in the form of buying higher quality seed potatoes, fertilizers or other agro-chemicals. It could also require investments in time in the form of integrated management of bacterial wilt or improved on-farm seed potato selection. Farmers will be reluctant to invest these resources required for adopting innovative technology that could improve their productivity when there is no clear market incentive to do so. Certainly for Nyandarua in Kenya, but basically for all districts, technological innovation by farmers would be helped by reducing marketing insecurities. As such, parallel technical and marketing interventions would increase the probability of success.

2.7.2 Knowledge development and information exchange

Knowledge development and information sharing are essential for the achievement of potato system improvement through technological innovation, and these are key functions of the innovation system. In the stakeholder workshops in the three countries, the interaction between actors managing potato information was found inadequate and identified as a key area for improvement.

Douthwaite (2002) looks at innovation from the perspective of fostering technological change, and proposes the "learning selection approach", an evolutionary approach towards technological development in which bright ideas are tested and adapted until they are "fit" enough to become useful. The technological innovations "surviving" this process will have become adapted to the environment in which they have been developed. As such imperfections in the existing system are "fixed" with adapted technology.

Other authors put more emphasis on the co-evolution of the technology and its environment, under the assumption that both are dynamic and will require more or less adaptation to facilitate positive change (Biggs and Smith 1998; Leeuwis 2000; Campilan 2002; Elzen and Wieczorek 2005; Geels 2005). Depending on the actual technology, innovation requires to a larger or lesser extent changes in the interrelations and formal and informal rules of conduct between stakeholders. Successful innovation thus depends on moulding and adaptation of both the technology and the environment in which it operates.

The stimulation of rural development, or in other words the 'moulding and adaptation process', is best viewed from a negotiation and learning perspective, rather than as a well structured planning and decision making process (Leeuwis 2000). The processes of network building, social learning and negotiation deserve particular support as they can catalyse system innovation (Leeuwis 2004). Similarly, Campilan (2002) concluded from empirical cases in Nepal that a social learning process is required for successful potato integrated disease management (IDM). In this social learning process, stakeholders jointly define the problem situation, design technical interventions, set up the corresponding social arrangements and learn to manage the links between these social and technical components.

Whether one focuses primarily on the evolution of technology, or on the co-evolution of technology and its environment, an understanding of the current dynamics of the system is required. It is essential to understand the existing potato system and its interactions with "niche-innovation" (new technology) and the wider environment or "socio-technical landscape" (Geels and Schot 2007). Innovation through a negotiation and learning process can only be initiated after it has been assessed whether the vital pre-conditions are met, which include institutional manoeuvering space, a sense of urgency of change among stakeholders, and a basic level of trust between stakeholders (Leeuwis 2004).

Engel (1997) proposes to analyse systems from a stakeholder interaction perspective, with particular focus on the knowledge and information system. Following Engel (1997) the "AKIS-potato" could be defined as a group of individuals, public organizations (governmental and non-governmental) and the private sector that exchange information and knowledge related to potato management, processing and trade. Understanding this system, its components and the way in which they interact is the essential first step for a more efficient innovation system (Lundvall et al. 2002; Hall et al. 2003).

2.7.3 Quantification of information sources of potato producers

Through the knowledge and information survey potato farmers were asked to indicate the initial sources of information on potato production technologies and marketing. Table 2.9, Table 2.10 and Table 2.11 show from where the sampled potato producers indicated they obtained information on different aspects of potato production and marketing in Kenya, Uganda and Ethiopia respectively.

Table 2.9: Summary of farmer information sources in production and marketing of potato in Nyandarua and Bomet districts in Kenya, 2004.

Farming[a] practice	Most important source of information (%)								n[b]
	Own experience	Family member	Farmer own community	Extension / NGO	Research	Publication / media	Private sector	Other	
Potato varieties	12.6 II	5.0	73.4 I	6.0 III	1.5	1.0	0.0	0.5	199
Seed potato selection	48.5 I	6.8	22.0 II	21.1 III	0.0	1.7	0.0	0.0	355
Soil fertility management	58.9 I	6.9	13.1 II	18.3 III	0.0	1.7	0.6	0.6	175
General crop husbandry	54.9 I	11.7	13.1 III	19.4 II	0.2	0.2	0.4	0.2	557
Post harvest handling	48.7 I	14.8	19.6 II	16.9 III	0.0	0.0	0.0	0.0	189
Marketing	41.4 I	15.2 III	23.2 II	11.1	0.0	7.1	2.0	0.0	99
Crop protection	29.0 I	7.0	27.0 II	20.0 III	2.0	2.0	13.0	0.0	100

[a] Bold Roman numbers highlight the three most mentioned information sources for each farming practice
[b] Within each farming practice category, different topics were included (not presented). The farmers were asked to only mention the single, most important source of information for these topics. n refers to the total number of valid responses on the different topics within a farming practice category. Total number of respondents = 97.
Source: knowledge and information survey

Table 2.10: Summary of farmer information sources for production and marketing of potato in Kabale district, Uganda, 2005

Farming[a] practice	Most important source of information (%)							n[b]
	Own experience	Family member	Farmer own community	Extension / NGO	Research	Potato traders	Other	
Potato varieties	15.9 II	11.3 III	69.2 I	0.0	2.6	0.0	0.5	195
Seed potato selection	46.8 I	0.0	24.7 III	25.9 II	2.5	0.0	0.0	158
Soil fertility management	46.2 I	8.9 III	8.9 III	32.3 II	3.6	0.0	0.6	303
General crop husbandry	25.9 I	20.1 III	15.6	34.6 II	3.8	0.0	0.2	680
Post harvest handling	43.9 I	0.6	29.0 II	23.9 III	2.5	0.0	0.0	314
Marketing	26.8 II	3.3	60.7 I	1.1	0.0	8.2	0.0	183
Crop protection	14.4 III	4.2	33.3 III	43.2 II	4.8	0.0	0.0	333

[a] Bold Roman numbers highlight the three most mentioned information sources for each farming practice
[b] Within each farming practice category, different topics were included (not presented). The farmers were asked to only mention the single, most important source of information for these topics. n refers to the total number of valid responses on the different topics within a farming practice category. Total number of respondents = 211.
Source: knowledge and information survey

Table 2.11: Summary of farmer information sources for production and marketing of potato in Ethiopia, North Shewa, South Shewa and East Hararghe zones, 2004

Farming[a] practice	Most important source of information (%)						n[b]
	Own experience	Family member	Farmer own community	Extension / NGO	Research	Potato traders	
Potato varieties	3.0	14.8 III	58.7 I	17.9 II	5.5	0.0	797
Seed potato selection	57.3 I	24.3 II	15.6[b] III	2.8	-	-	1,109
Soil fertility management	63.7 I	15.3 II	11.1 III	6.3	2.9	0.7	1,128
General crop husbandry	59.2 I	16.2 II	16.1 III	4.8	3.4	0.4	3,203
Post harvest handling	62.9 I	17.0 II	9.2 II	5.9	4.2	0.7	707
Marketing	70.4 I	20.0 II	0.0 I	4.5	5.1 III	0.0	375
Crop protection	28.4 II	14.7	33.7 I	17.9 III	5.3	0.0	638

[a] Bold Roman numbers highlight the three most mentioned information sources for each farming practice
[b] Includes research and traders
[c] Within each farming practice category, different topics were included (not presented). The farmers were asked to only mention the single, most important source of information for these topics. n refers to the total number of valid responses on the different topics within a farming practice category. Total number of respondents = 646.
Source: knowledge and information survey

Information on crop husbandry, seed potato selection and post harvest handling, which entail most of the routine farming operations, was considered by farmers to be developed through own experience. Apparently many farmers do not regard this knowledge as derived from an identifiable source of information, but rather evolved through experience of the user. A considerable proportion of farmers in Uganda and Kenya indicated the public agricultural extension service, non-governmental organizations in the field of agricultural development (hereafter called NGOs) or events organized by these actors as a principal source of information on practices related to seed potato selection, soil fertility management, general crop husbandry and post harvest handling. In comparison, in Ethiopia the combined public and non-governmental extension actors play a more marginal role as a source of information on these farming practices.

The majority of farmers indicated to have obtained first information about the potato cultivars that they grow from fellow farmers within the local community. Family members were a second source of information on potato variety. Extension and NGOs only play an important role regarding information on cultivars in Ethiopia, while the research organizations were mentioned sporadically as a source of information on potato cultivars in each country. Potato traders seemed not to play any role in information on new potato cultivars.

For information on crop protection farmers relied less on their own experience and the role of extension was more important here than in other categories. Also research was mentioned more frequently as a source of information than in general crop husbandry practices. Only in Kenya the input suppliers were mentioned as a source of information on crop protection, which indicates that the input market for potato is relatively less developed than in other parts of the world.

In Kenya commodity prices at wholesale markets are published through radio and newspapers, while farmers also received information from neighbours and family members. In Uganda, neighbouring farmers were the most important source of information on potato marketing and farmers in Ethiopia mostly relied on their own experience and that of their family members to source information for potato marketing, suggesting that there is a limited role of government or private services in the provision of this type of information to farmers.

The general observation can be made that the own farming community and family members are important sources of information on potato farming practices and techniques. Farmer to farmer information flow outside the own community or family on the contrary is almost absent. The major actor facilitating access to potato related information from outside the local community or family is the agricultural extension. The direct contribution of research organizations to the knowledge and practices of potato farmers is modest but measurable. Considering their limited outreach, research organization are required to work closely with NGO and public extension services to assure technological innovations get a reasonable chance to evolve outside the research organizations' chosen pilot testing areas. Publications and media play a very marginal role in information exchange and knowledge on farming practices. Even for technologies that require inputs, such as soil fertility management and crop protection, the role of the supplier as a source of information is limited. Potato traders were not important as a source of information on potato farming, cultivars nor marketing.

2.7.4 Important actors in the potato related innovation systems

During the stakeholder workshops the important actors and their roles in the potato innovation system were identified. The Kenyan potato related innovation system was characterized by the presence of the public extension service of the ministry of agriculture as the sole institutional provider of information at the farmer level. The national research organization, international research institute as well as a development donor collaborated with the ministry of agriculture in initiatives in the potato sector. There is a specific body for seed potato certification and several actors specifically intervening in the seed potato system. Other innovation system actors identified by the workshop participants who are playing less intensive and visible roles are potato traders, input suppliers, universities and local governments. Potato farmer organization is limited to self-help groups in the potato growing zones. A national potato farmer association is attempting to establish and earn the mandate and recognition of the potato farmers as the organization representing them in lobbying and marketing. The producers, however, expressed reluctance and suspicion towards attempts to organize the farmers as a result of a long tradition with such organizations being used for the self interest of its officials, rather than the interest of its members.

In Uganda, the potato knowledge and information system is characterized by a public extension service that is being replaced by a government funded but private delivered agricultural extension program (National Agricultural Advisory Services or NAADS) (Benin et al. 2007). Through a decentralization process local governments have become important actors in the agricultural system as indicated by the workshop participants. During the workshop several NGOs were identified that operate in the potato production zones of south western Uganda, in different alliances with each other and the national and international research organizations, the private extension service delivery NAADS, the conventional extension services and the local government system (Gildemacher et al. 2009a). Consequently, there is a relatively strong interaction between different actors in the potato system. There is no formal seed potato certification system, but there are several actors specifically intervening in the seed potato system. As in Kenya, potato traders and agro-input suppliers were hardly recognized as key actors in the knowledge and information system. Farmer organization is, as in Kenya, largely limited to self-help groups.

The Ethiopian system was characterized by a centralized organization of both research and public extension. Non-governmental intervention was coordinated at a macro level by the central government. In spite of this central coordination, there was limited interaction between agricultural research, public extension and NGOs, which was identified as a major obstacle for potato system innovation during the stakeholder workshop. In different potato growing areas, universities are active in the potato sector but also not always in coordination with the national research organization programs. Farmer organization is limited although all farmers are by default members of peasant associations, the lowest

Table 2.12: Identification of system failure in the potato related innovation system in Eastern Africa

Rules (System failures) \ Actors	International research (CIP)	National research organizations	Extension (Public, private and NGO)	Potato growers	Private sector
Infrastructural failure (constraints requiring major investments)	• Low staff numbers • No research infrastructure Africa • No seed potato infrastructure	• Low staff numbers • Low mobility • Limited web access • Insufficient research infrastructure • Limited seed potato infrastructure	• Low staff numbers • Low mobility • No web access	• Bad roads • Poor marketing infrastructure • No storage capacity • Poor availability inputs (Ug & Et) • No credit facilities • Limited processing	• Bad roads • Poor marketing infrastructure • No credit facilities
Hard institutional failure (formal, written, consciously created rules of conduct and interaction)	• Limited responsiveness resulting from funding mechanisms • Narrow interpretation of research mandate	• Rigid administrative systems • No cross country germplasm exchange • Complicated variety release (Ke)	• Limited responsiveness due to rigid mandates and intervention areas • Centralized resource flows	• Complicated seed potato certification (Ke) • Lack of farmer organization	• Lack of quality standards • Lack of professional body
Soft institutional failure (informal rules of conduct and interaction)	• Hierarchical perception of roles of actors • A high tech bias • Large farmer bias • Fast staff turnover • Low priority for capacity building partners • Low priority communication of research results	• Hierarchical perception of roles actors • High tech bias • Large farmer bias • Acceptance sub-optimal functioning organization • No resource sharing • No performance based staff rewarding • Limited international orientation and innovative thinking • Low priority communication of research results	• Hierarchical perception of roles actors • High tech bias • Large farmer bias • No resource sharing • No performance based rewarding (public ext.) • Limited faith in farmer capacities • Passive role in research • Fast staff turnover (NGO) • Short planning horizon (NGO)	• Acceptance poor bargaining power • Acceptance current production and knowledge level • Mistrust of farmer organizations (Ke, Et) • Limited trust extension service • Passive information receivers • Lack of incentives for higher quality production	• Acceptance low quality potatoes • Acceptance imperfect marketing systems
Strong network failure (strong ties between actors hampering interaction with outside actors)	• Single focus on national research organizations as partners	• Single focus on public extension as "messenger" of technological innovation • Single focus on CIP for delivery of ideas, resources and new technology	• Single focus on national research as a source of new information • No pro-active project acquisition or search for partnership	• Single focus of farmers on extension for clean seed and information.	• Single focus on traders and middlemen as source of potatoes
Weak network failure (limited interaction between actors)	• No interaction with private sector • Poor grassroots connection • Limited interaction with public extension	• No interaction with private sector • Localized grassroots and public extension interaction	• No interaction with private sector • Limited and passive interaction with research • Limited interaction with selective group of farmers • No interaction with private sector	• No interaction with private sector • Limited interaction outside of the community • No farmer organization	• No interaction with producers • No interaction with extension • No interaction with research
Human capacity failure (essential skills missing for playing an effective role in the innovation system)	• Limited capacity for hands-on on-the-job capacity building of collaborators • Limited capacity to design and test communication strategies	• No innovative thinking • Low multidisciplinary research capacity • No skills for packaging and communicating research skills • Limited stakeholder interaction management capacity • No impact assessment / evaluation	• Limited capacity in participatory approaches / adult learning / group dynamics • No capacity to develop training materials / methods • Limited stakeholder interaction management capacity • Low organizational capacity • No impact assessment / evaluation	• Very low education levels • Illiteracy • Limited experience and incentive to innovate • No organizational skills	• No research capacity • No skills in participatory interaction with smallholder farmers

Ethiopian administrative unit. Potato traders and agro-input suppliers are few and judging from the stakeholder workshop outcomes as well as from Table 3.11, they hardly play a role in the potato knowledge and information system. The NGO and government bodies represented in the stakeholder meeting indicated to prioritize the organization of farmer unions and associations, but indicated that this is complicated as they have to overcome farmer suspicion of such initiatives as a result of forced farmer organization under totalitarian rule in the recent past.

2.8 Potato innovation system constraints identified

Klein Woolthuis et al. (2005) developed a "system failure framework" as a tool for policy recommendation in the industrialized world. The same approach was used for synthesizing the insights in the potato system in Kenya, Uganda and Ethiopia. Klein Woolthuis et al. (2005) are distinguishing six different categories of system failure. These are infrastructure failure, hard institutional failure, soft institutional failure, strong network failure, weak network failure and capabilities failure. Infrastructural failures are constraints requiring major investments, with returns in the long term, that can not be made by the actors of the system independently. Hard institutional failure refers to rules and regulations, or the lack of them, hampering innovation, whereas soft institutional failure refers to unwritten rules, or "the way business is done". Strong network failure refers to actors "locked" into their relationship, which blocks new ideas from outside and prohibits other potentially fruitful collaborations. Weak network failure refers to a situation where actors are not well connected and "fruitful cycles of learning and innovation may be prevented" (Woolthuis et al. 2005). Finally capabilities failure points to the lack of technical and organizational capacity within the system to adapt to and manage new technology (Woolthuis et al. 2005).

In Table 2.12, the potato sector failures are presented according to these categories, using the information about the innovation system obtained through the stakeholder workshops, the surveys and expert opinion of other innovation system actors. The different types of system failures are cross tabulated against the relevant innovation system actors, to provide a structured insight in the functioning of the system. Table 2.12 structures the shortcomings of the potato innovation system, and provides insight in the types of failures of the different important actors in the system. The failures per actor are briefly discussed below.

2.8.1 International potato center

The weaknesses of CIP lie in the limited staff with a high turnover and severe limitations in research infrastructure in Africa. The strategy of CIP to deal with the low staff numbers is a strong focus on the national research organizations as partners, while there is fairly limited interaction with the private sector and the public extension services. CIP shares a high tech focus with the national research organizations which promotes a bias towards larger successful farmers. The limited staff numbers also result in relatively weak grassroots connection of the institute within Africa. As a result of a rigid description and interpretation of the boundaries of its mandate, limited priority is given to capacity building of the other actors in the system and to the appropriate packaging and communication of research results (Table 2.12).

2.8.2 National research organizations

Although the national research organizations in the three countries differ significantly, there are commonalities in their failures in the system. Insufficient research facilities and seed potato production facilities are serious constraints according to the national researchers. In Kenya and Uganda communication is reasonably good, although researchers rely largely on private mobile phones and internet cafes for their communication. In Ethiopia access to mobile phones and internet cafes was less developed. Mobility is restricted due to limited funds for vehicle replacement, maintenance and use, resulting in complaints by both farmers and extension workers that researchers do not respond quickly to their needs and are not visible in the field. The functioning of the national research organizations is further complicated by rigid administrative systems of the institutions as well as from donors that complicate the access to and effective spending of the limited available funds. The functioning of the research organizations is also hindered by a lack of performance based rewarding of staff (Table 2.12).

Regional research collaboration exists, but potato researchers in the region indicate they experience severe constraints in communication, as well as in mobility within the region. Collaboration in crop improvement is hampered by rigid rules regarding cross border exchange of germplasm and seed. Different regulations for potato variety release procedures in the three countries complicate the regional release and utilization of same cultivars across the region. In Kenya variety release regulations are overcomplicated.

The national research organizations are also biased towards high-tech problem-solving approaches. Additionally, interventions and research collaborations tend to target more successful and affluent farmers. CIP is the major source of innovative potato technologies for testing and adaptation in the national potato farming systems. Limited priority is given to the effective and appropriate communication of research results to end-users. Within the national research organizations there is a lack of capacity to facilitate stakeholder interactions, execute multi-disciplinary research and perform impact assessments and evaluations of interventions.

Sharing of resources with extension partners is rare, except in few specific projects. Collaboration with extension partners and farmers is mostly *ad hoc* and local, without a vision on maximizing impact of innovative technology. Very limited interaction exists between the national research organizations and the private sector, partly because the private sector is poorly developed and partly as a result of mistrust.

2.8.3 Agricultural extension actors

The complex of agricultural extension actors includes the public agricultural extension services, private agricultural extension service providers and non-governmental organizations all engaged in agricultural extension activities. This is a heterogeneous group with different limitations in their roles in the potato system in the three countries. The limitations were identified in the stakeholder workshops and through the consultation with the actors.

The public extension service suffers from low staff numbers and the mobility of the limited staff is very poor as they lack transport means. In all three countries extension workers lack communication means and have no access to professional information. In Kenya and Uganda the functioning of extension workers is further complicated by continuous reorganization of administrative structures. In Uganda the public extension services have been largely dismantled in favor of the contracting of "private extension service providers" under the NAADS program (Benin et al. 2007). In all the three countries farmers complained of limited access to the extension staff. When asked about the major problems in sourcing information farmers most often mentioned the lack of presence in the field of extension workers.

All extension actors are limited by the specific mandates they have through project financing, mostly so the private service providers in Uganda and NGOs, who are bound to contracts and project defined intervention areas. This tended to leave out many potential beneficiaries. The system also tended to the same bias as researcher organizations of favoring affluent or successful farmers.

The public extension services and private service providers look unilaterally to the national research organizations for solutions to problems indicated by farmers, and indicated in the workshops they are disappointed in the response time of research. Like the other actors the extension services lack contact with private sector partners and there is mistrust between them and potato traders and input suppliers. In both Kenya and Ethiopia extension workers explicitly indicated that input suppliers cheat potato farmers. Furthermore they felt that input suppliers do not have the skills to play a role as sources of information on potato production.

In general the capacity and experience of the agricultural extension workers in participatory approaches, adult learning techniques and group dynamics was low. Like the national research organizations the extension organizations lack the capacity to develop appropriate training methods and materials. They have low organizational capacity and limited experience in managing stakeholder interaction. With the exception of non-governmental organizations no impact assessment and project evaluation is practiced routinely.

2.8.4 Potato growers

The interests of potato growers are not represented by any professional body in the three countries. Farmers in the region are not organized beyond village-based self-help groups. As a result of this lack of organization potato producers at a higher level they importantly lack in clout in potato marketing as well as in the general potato related innovation system.

During the stakeholder workshops farmers indicated that poor rural road infrastructure and the poor marketing infrastructure aversely affect the potato market. In absence of on-farm ware potato storage facilities, market information and farmer organizations farmers are not able to bargain for better prices with middle men. Marketing infrastructure is limited in all three countries, but most particularly in Ethiopia, where farmers transport and sell their potato produce piecemeal to buyers at local village markets (Table 2.6). At the same time, researchers and NGO workers indicated the absence of wholesale markets for potatoes in the capital in Ethiopia.

In Kenya and Uganda there are traders who collect and sell potato to wholesalers in urban markets. There are hardly any alternatives to farmers than to market their produce for the fresh potato market, especially in Ethiopia and Uganda. In Kenya, a limited processing industry for crisps and an infant industry for frozen chips exist (Hoeffler and Maingi 2005). Potato farmers lack access to credit facilities in all three countries. This was put high on the list of priorities by farmers in the Kenyan stakeholder workshops. In Uganda access to inputs is limited for farmers as both fungicides and fertilizers are hardly available outside the main town centers. Farmers in Ethiopia indicated that unavailability of fungicides when it is most needed in rural areas is a major constraint.

Access to high quality seed potatoes is cumbersome for potato farmers in all three countries, and mentioned by producers and extension workers as a major bottleneck for potato production. In the stakeholder workshops in both Kenya and Ethiopia producers and extension workers blamed the inadequate production and distribution of high quality seed potatoes squarely on the research organizations.

Direct contact between producers and retailers or processors of potatoes is limited to few cases initiated through development projects at considerable effort. The existing marketing situation does not provide much incentive for innovation. Farmers accept the current low levels in knowledge and production and are in general not pro-active in seeking collaboration with extension services which they do not hold in very high regard, with the notable exception of NGOs who have been distributing improved potato cultivars and high quality seed.

2.8.5 Private sector

Although potato producers could also be considered private entrepreneurs, the private sector is here understood to be those actors that commercially deliver goods and services to the potato production chain, other than potato production. This comprises agro-input and output marketing and processing.

Most importantly the processing and retailing sector indicates to suffer from irregular and poor quality supply of potatoes. There are no standards for potato grading and packaging or even variety names on which potato wholesalers, retailers and processors could rely when judging quality. Furthermore representatives of the output marketing and processing sector in Kenya identified the lack of potato storage facilities as an important problem. The private sector actors are not organized in a professional body representing their interests that could function as a first contact point for the other actors that may desire to interact.

The private sector largely relies on informal brokering, transporting and wholesaling systems that exist in the countries, with limited interference to optimize the commodity flow. The majority of private sector actors has no interaction with potato producers, with the exception of input retailers and field level potato brokers. Few private sector actors interact with research and extension organizations, with the exception of few specific individuals.

2.9 Discussion and conclusions

2.9.1 Potato production and marketing

This research has shown that potato production provides smallholder farmers in the East African highlands with a profitable exploitation of their scarce land and capital, as most farmers get a higher return on their labour by growing potatoes than they would get by hiring out their labour. Considering the limitations to increasing the share of arable land to potatoes, especially in Kenya and Uganda, an increase in productivity per unit area is required to off-set rising demand for potatoes in Kenya, Uganda and Ethiopia.

Four technical intervention areas were identified which could lead to improved productivity as a result of technological innovation. These included: seed potato quality management, bacterial wilt control, late blight control and soil fertility management.

2.9.2 Potato innovation system

Improving productivity alone will not result in a dramatic improvement of profitability of potato production. Marketing was a prime concern of potato farmers, and logically a major driver of decisions. Technical innovation in conjunction with marketing intervention would increase the probability of successful intervention in the smallholder potato system in Kenya, Uganda and Ethiopia. The current innovation system has internal flaws that could hamper development of solutions to the production and marketing constraints of the potato sector.

An important feature of the potato related innovation system is the combination of soft institutional and strong and weak network failure in the relationship between the research organizations, extension services and farmers in the system (Table 2.12). There is "soft institutional failure" in the sense that each actor has a fairly conservative narrow interpretation of its and others' roles, along the linear transfer of technology model, already criticized by Robert Chambers (1983). Chambers identifies the "normal professionalism" of agricultural scientists as the cause of the domination of the ineffective linear transfer of technology model. Hierarchical thinking about actor roles, large farmer bias and the related high tech bias are elements of this "normal professionalism". From the system analysis in the region, it is evident that not only researchers and extension workers but also potato farmers suffer from hierarchical thinking about roles of actors.

The potato related innovation system suffers simultaneously from "strong network failure" and "weak network failure". On the one hand research, public and NGO extension and farmers do not look beyond their horizons for solutions to problems. Simultaneously there is weak network failure in the form of the lack of productive partnerships between private sector partners and research, extension services and producers. This combined with the static interpretation of each others roles limits the chances of system innovation through new ideas.

To overcome network failures, facilitation of stakeholder interaction through a potato platform was identified as priority in the three stakeholder meetings (Gildemacher et al. 2009a). Impartial intermediaries could be considered to improve the innovation system interaction and coordination. These have been shown to be well positioned to restore the functioning of an innovation system (Klerkx and Leeuwis 2008). A difference with the system described by Klerkx and Leeuwis (2008) is however that the potato innovation systems in the three countries are hardly privatized, and there is a collaborative rather than a competitive sentiment between the most important potato system actors. As such an impartial intermediary may not be essential and another option could be for one of the actors to take on the role as system broker, and champion the potato innovation system improvement.

There is a clear need for smallholder potato producers to unite themselves to raise their clout in both knowledge demand articulation as well as in marketing. Their organization at a higher level is pivotal to the success of improved innovation system functioning. Considering the low level of training of the average potato producer the organizational capacity of the producers is limited and effective farmer organization is thus unlikely to be initiated without professional support. A starting point for formation of an umbrella potato farmers association could be the currently the current rudimentary, village-level self help groups present in all three countries.

In spite of the widely acknowledged criticism on the linear transfer of technology model, extension services, in whichever form, are an important and necessary link between end-users and agricultural research, both for demand articulation as well as for the communication of innovative technology. The extension services need to get involved in technical innovation efforts to tackle the four technical constraints to productivity increase. The other innovation system actors should take into consideration the limited mobility and the low level of training of field workers.

Research and extension partners realised they lack the capacity to translate research insights and pilot experiences of research and development partners into generically useful training materials that could support and facilitate a wider adoption of innovations. The potato sector in the three countries therefore should not only focus on its technical research capacity, but simultaneously develop skills and commit resources to improvement of the service delivery to potato producers.

References

Benin S, Nkonya E, Okecho G, Pender J, Nahdy S, Mugarura S, Kayobyo G (2007) Assessing the impact of the National Agricultural Advisory Services (NAADS) in the Uganda rural livelihoods. International Food Policy Research Institute (IFPRI), Addis Abeba.

Biggs S, Matsaert H (1999) An actor-oriented approach for strengthening research and development capabilities in natural resource systems. Public Administration and Development 19:231-262

Biggs S, Smith G (1998) Beyond methodologies: Coalition-building for participatory technology development. World Development 26:239-248

Campilan D (2002) Linking social and technical components of innovation through social learning. The case of potato disease management in Nepal. In: Leeuwis C, Pyburn R, Röling N (eds) Wheelbarrows full of frogs: social learning in rural resource management. Koninklijke Van Gorcum, Assen, pp 135-146

Chambers R (1983) Rural development: putting the last first. Longman, London [etc.]

Douthwaite B (2002) Enabling innovation: a practical guide to understanding and fostering technological change. Zed Books [etc.], London [etc.]

Elzen B, Wieczorek A (2005) Transitions towards sustainability through system innovation. Technological Forecasting and Social Change 72:651-661

Engel PGH (1997) The social organization of innovation: a focus on stakeholder interaction. Royal Tropical Institute, Amsterdam

Fuglie KO (2007) Priorities for potato research in developing countries: results of a survey. American journal of potato research 84:353-365

Geels FW (2005) Processes and patterns in transitions and system innovations: Refining the co-evolutionary multi-level perspective. Technological Forecasting and Social Change 72:681-696

Geels FW, Schot J (2007) Typology of sociotechnical transition pathways. Research Policy 36:399-417

Gildemacher P, Demo P, Kinyae P, Nyongesa M, Mundia P (2007a) Selecting the best plants to improve seed potato. LEISA Magazine 23:10-11

Gildemacher PR, Landeo J, Kakuhenzire R, Wagoire W, Nyongesa M, Tessera M, Bouwe N, Bararyenya A, Hakazimana B, Senkesha N, Gashabuka E, Muhinyuza J, Forbes G, Lemaga B (2007b) How to integrate resistant variety selection and spray regime research for IPM of potato late blight in Eastern and Central Africa. 7th triennial African Potato Association conference. African Potato Association, Alexandria, Egypt, pp 84-92

Gildemacher PR, Maina P, Nyongesa M, Kinyae P, Gebremedhin W, Lema Y, Damene B, Shiferaw T, Kakuhenzire R, Kashaija I, Musoke C, Mudiope J, Kahiu I, Ortiz O (2009a) Participatory Analysis of the Potato Knowledge and Information System in Ethiopia, Kenya and Uganda. In: Sanginga, P.C., Waters-Bayer, A., Kaaria, S., Njuki, J., Wettasinha, C. (eds.) Innovation Africa: Enriching farmers' livelihoods. Sterling: Earthscan. Pp 153-167.

Gildemacher, P.R., Demo, P., Barker, I., Kaguongo, W., Woldegiorgis, G., Wagoire, W.W., Wakahiu, M., Leeuwis, C., and Struik, P.C. 2009b. A description of seed potato systems in Kenya, Uganda and Ethiopia. Amercian Journal of Potato Research. Accepted for publication.

Hall A, Rasheed Sulaiman V, Clark N, Yoganand B (2003) From measuring impact to learning institutional lessons: an innovation systems perspective on improving the management of international agricultural research. Agricultural Systems 78:213-241

Hoeffler H, Maingi G (2005) Rural-urban linkages in practice: promoting agricultural value chains. Entwicklung & Ländlicher Raum 2005:3-4

Klerkx L, Leeuwis C (2008) Matching demand and supply in the agricultural knowledge infrastructure: Experiences with innovation intermediaries. Food Policy 33:260-276

Leeuwis C (2000) Reconceptualizing participation for sustainable rural development: towards a negotiation approach. Development and Change 31:931-959

Leeuwis C (2004) Fields of conflict and castles in the air. Some thoughts and observations on the role of communication in public sphere innovation processes. The Journal of Agricultural Education and Extension 10:63 - 76

Lekasi JK, Tanner JC, Kimani SK, Harris PJC (2003) Cattle manure quality in Maragua District, Central Kenya: effect of management practices and development of simple methods of assessment. Agriculture, Ecosystems and Environment 94:289-298

Lemaga B, Kakuhenzire R, Kassa B, Ewell P, Priou S (2005) Integrated control of potato bacterial wilt in Eastern Africa: the experience of African highlands initiative. In: Allen C, Prior P, Hayward A (eds) Bacterial Wilt Disease and the *Ralstonia solanacearum* Species Complex. APS Press, St. Paul, pp 145-157

Low JW (1997) Potato in southwest Uganda: threats to sustainable production. African Crop Science Journal 5:395-412

Lundvall B-Å, Johnson B, Andersen ES, Dalum B (2002) National systems of production, innovation and competence building. Research Policy 31:213-231

Nyankanga RO, Wien HC, Olanya OM, Ojiambo PS (2004) Farmers' cultural practices and management of potato late blight in Kenya highlands: implications for development of integrated disease management. International Journal of Pest Management 50:135-144

Scott GJ, Rosegrant MW, Ringler C (2000) Global projections for root and tuber crops to the year 2020. Food Policy 25:561-597

Struik PC, Wiersema SG (1999) Seed potato technology. Wageningen University Press, Wageningen

Turkensteen LJ (1987) Survey of diseases and pests in Africa: fungal and bacterial diseases. Acta Horticulturae 213: 151-159

Vanlauwe B, Giller KE (2006) Popular myths around soil fertility management in sub-Saharan Africa. Agriculture, Ecosystems and Environment 116:34-46

Wakahiu MW, Gildemacher PR, Kinyua ZM, Kabira JN, Kimenju AW, Mutitu EW (2007) Occurrence of potato bacterial wilt caused by Ralstonia solanacearum in Kenya and opportunities for intervention. 7th triennial African Potato Association Conference. African Potato Association, Alexandria, Egypt, pp 267-271

Woolthuis RK, Lankhuizen M, Gilsing V (2005) A system failure framework for innovation policy design. Technovation 25:609-619

3 Seed potato systems in East Africa: description and opportunities for improvement

Peter R. Gildemacher[a,h], Paul Demo[b], Ian Barker[c], Wachira Kaguongo[a], Gebremedhin Woldegiorgis[d], William W. Wagoire[e], Mercy Wakahiu[f], Cees Leeuwis[g], Paul C. Struik[g].

[a] International Potato Center, Nairobi, Kenya.
[b] International Potato Center, Lilongwe, Malawi
[c] International Potato Center, Lima, Peru
[d] Ethiopian Institute for Agricultural Research, Holetta, Ethiopia
[e] Kachwekano Zonal Agricultural Research and Development Institute, Kabale, Uganda
[f] Kenya Agricultural Research Institute, Tigoni, Kenya
[g] Wageningen University and Research Centre, Wageningen, The Netherlands
[h] Royal Tropical Institute, Amsterdam, The Netherlands

Published in: American Journal of Potato Research (2009) 86:373-382.

Abstract

Seed potato systems in East Africa are described and opportunities for improvement identified on the basis of interviews with potato producers in Kenya, Uganda and Ethiopia, and an assessment of *Ralstonia solanacearum* and virus disease levels in Kenya. 3% of seed potato sold in Kenyan markets was virus free. *Ralstonia solanacearum* was found in 74% of potato farms. Less than 5% of the farmers interviewed source seed potato from specialized seed growers. Over 50% rely entirely on farm-saved seed. Current seed potato prices justify this behavior. To improve the system, both the local and specialized chain need to be tackled simultaneously. To improve the local chain ware potato farmers require training on seed quality maintenance and managing bacterial wilt and viruses. Research into virus resistance and the effect of mixed virus infection on yield deserves attention. Private investment in seed potato production could increase volumes produced and reduce prices.

3.1 Introduction

In potato production the quality of seed potatoes planted is an important determinant of the final yield and quality (Struik and Wiersema, 1999). If farm saved seed potatoes are used for several cropping cycles, without renewing the seed lot from a reliable source, seed-borne diseases cause severe yield and quality losses. This process of yield loss over seasons of seed recycling is generally called degeneration, and can be attributed to the accumulation of seed borne diseases (Gildemacher et al., 2007).

Turkensteen (1987) identified bacterial wilt, caused by *Ralstonia solanacearum* and virus diseases caused by PVY and PLRV as the major seed borne potato diseases, but also mentioned soft rot (caused by *Erwinia chrysanthemi*), Fusarium wilt and dry rot (caused by *Fusarium solani*) and Verticilium wilt (caused by *Verticilium albo-atrum*) as economically important seed borne diseases. Gildemacher et al. (2008) consider virus diseases and potato bacterial wilt as the most important seed borne potato diseases in Eastern Africa. Low seed potato quality is believed to be one of the major yield reducing factors in potato production in Sub Sahara Africa (Fuglie, 2007). This contributes to the low average yields in Sub-Saharan African countries of around 8 Mg/ha on the continent compared to a world average of 16 Mg/ha (FAO, 2008). Serious yield losses can be expected as a result of high infection rates with potato viruses (Reestman, 1970).

The problem of seed degeneration has been solved in the Northern potato producing countries through specialized seed potato producers (hereafter called seed growers) who multiply seed potatoes from basic pathogen free starter seed. Consumption potato producers (hereafter called ware growers) maintain maximum production potential over the seasons by replacing their seed potato stock each season or at least frequently with high quality seed potatoes from a seed grower. This keeps the virus pressure in the entire cropping system low. This so-called "flush out" system is practised in more advanced potato cropping systems in the Western world (Struik and Wiersema, 1999).

Different systems of seed potato multiplication have been initiated in potato growing developing countries in the world, including Sub-Sahara Africa (Monares, 1987; Potts and Nikura, 1987; Crissman et al., 1993), resulting in different scales of success. In spite of the many efforts with regards to seed potato system improvement, potato farmers in Kenya, Uganda and Ethiopia still identified seed potato quality as their major concern within their potato production system and it was prioritized as an important technical intervention area to improve smallholder potato profitability (Gildemacher et al., 2006).

In this paper the current status of the seed potato systems in Kenya, Uganda and Ethiopia is described on the basis of surveys of potato producers to assess their seed potato production practices. Furthermore an assessment was made of the level of potato viruses in the general seed stock available to farmers in Kenya, and the severity of bacterial wilt in Kenyan potato fields. On the basis of the data opportunities for seed system improvement in Kenya, Uganda and Ethiopia are discussed.

3.2 Materials and methods

3.2.1 Quantification of the importance of seed borne diseases

To assess the importance of seed borne diseases, specific surveys were conducted in Kenya. Seed potatoes sold on rural markets were assessed via a survey for their level of infection with major potato diseases. With another survey, the level of bacterial wilt (*Ralstonia solanacearum*) in potato fields in major potato growing areas in Kenya was quantified.

Potato virus survey in Kenya

In September 2006 seed potatoes were sampled randomly in batches of 20 from four vendors in each of 11 rural markets, covering Nakuru, Nyandarua, Nyeri, Laikipia, Meru Central, Muranga and Kirinyaga districts, thus representing the major potato production zones of Kenya. Samples were carried to the laboratory in paper sample bags and stored until sprouting. Then Double Antibody Sandwich ELISA was conducted against four important potato viruses, PVY, PVX, PLRV and PVA, on the single tuber samples, using DAS ELISA from CIP, Lima (Salazar and Jayasinghe, 2002).

Bacterial wilt survey in Kenya

Three major potato growing areas in Kenya, Kiambu, Nyandarua and Bomet district, were purposefully chosen to represent different potato farming systems in Kenya. Thirty five, 32 and 34 potato fields were selected in Kiambu, Nyandarua and Bomet district respectively. The fields were randomly selected along rural unpaved roads at intervals of about 2.5–5 km. The severity of bacterial wilt was recorded in all potato fields of the visited farms by scoring the incidence of plants showing the characteristic symptoms of the disease. Fields with potato crops that had not yet developed a more or less closed canopy were not considered. The survey was conducted during the main rainy season of 2005, in the months of June, July and August when most farms had a well established potato crop.

3.2.2 Potato farming practices survey

A survey was conducted to document potato farming practices in Kenya, Uganda and Ethiopia. In each country first a rapid appraisal was executed, to allow for proper site selection and questionnaire development. The questionnaire was pre-tested in each country and adapted to local circumstances, and translated in Ethiopia. The survey was executed by specifically trained enumerators recruited locally within each country.

In Kenya, data collection took place between 10 and 29 October 2005. Meru Central and Nyandarua were selected as sample districts, as they were considered to best represent the whole of the Kenyan potato production system. A district is an administrative topographical unit, which is further sub-divided into divisions, locations and sub-locations, the latter being the smallest administrative unit. Six farmers were chosen randomly in half of the sub-locations within each location in the sampled districts, to get a satisfactory number of sample households. The sample households were randomly picked from a list of all farm households in the village, provided by a village elder. In total 251 farmers were successfully interviewed, 100 in Meru Central district and 151 in Nyandarua district.

Kabale and Kisoro districts were selected to represent the potato farming system in Uganda. Districts are sub-divided in counties, sub-counties, parishes and villages. All the 4 counties and the 25 potato producing sub-counties in Kabale and Kisoro districts were included in the study. One parish was randomly selected from each sub-county and one village was randomly selected within each parish. Six households were picked at random in each sampled village, from a list of households provided by a village elder, to assure a sufficiently large and representative sample. In all, 144 farmers out of 150 randomly selected were successfully interviewed.

In Ethiopia three major potato producing districts (woredas) were selected, Jeldu in West Shewa zone, Degem in North Shewa zone and Banja Shikudadin in Awi zone, as a cross section of potato production in the country. Within these districts six households were randomly selected within each kebele (village), resulting in 220 households that were successfully surveyed.

3.3 Results

3.3.1 Potato virus survey in Kenya

Table 3.1 shows the average infection rates for potato leaf roll virus (PLRV), potato virus Y (PVY), potato virus X (PVX) and potato virus A (PVA) in potatoes sold as seed potatoes on rural markets in Kenya. PLRV and PVY infection were highest. More than half of the sampled tubers were infected with PVX and slightly less than half of the potato tubers with PVA. Out of the total sample of 1000 tubers, only 27 tubers were found to be entirely free of these four viruses.

Table 3.1: Incidence of PLRV, PVY, PVX and PVA in seed potatoes sold at rural markets in Kenya, September 2006.

Market	District	Virus incidence levels (%)					
		PLRV	PVY	PVX	PVA	Virus free	Multiple infections
Elburgon	Nakuru	29	83	39	10	8	50
Kagio	Kirinyaga	68	91	83	56	0	96
Karatina	Nyeri	91	78	83	28	1	93
Kihingo	Laikipia	71	48	100	9	0	79
Mau Narok	Nakuru	61	83	30	15	9	68
Meru	Meru Central	91	58	70	40	1	84
Molo	Nakuru	49	70	64	14	6	66
Murang'a	Muranga	95	100	64	64	0	100
Nanyuki	Meru Central	96	100	55	65	0	98
Naru Moru	Nyeri	63	29	46	65	6	75
North Kinangop	Nyandarua	99	98	34	78	0	98
South Kinangop	Nyandarua	74	94	23	65	3	83
Grand mean		**74**	**77**	**57**	**42**	**3**	**82**

Source: own virus survey

3.3.2 Bacterial wilt survey in Kenya

The disease could be detected on the basis of outright symptoms in 74% of the farms. The incidence of wilting plants per farm was 0.78, 1.09 and 1.47% for Nyandarua, Kiambu and Bomet districts, respectively (Table 3.2).

Table 3.2: Bacterial wilt incidence and prevalence in potato farms in three districts in Kenya, 2005.

District	Mean wilt incidence (%)	Prevalence of bacterial wilt in farms (%)	No. of farms sampled
Nyandarua	0.78	69	32
Kiambu	1.09	63	35
Bomet	1.47	91	34
Mean	**1.12**	**74**	

3.3.3 Potato farming practices survey

Two seed potato production and marketing systems or chains could be identified, here called the specialized chain and the local chain. In the specialized chain seed growers produce and sell seed potatoes as a business, which includes non-certified commercial multiplication of starter seed, which others would classify as 'informal' seed potato multiplication (Crissman et al., 1993; Thiele, 1999; Tindimubona et al., 2000). The local chain can be characterized by the fact that seed potatoes are a by-product of ware potato production and are sold and traded locally, as also described by Crissman et al. (1993). Using the information of the surveys the importance of the different flows of seed potato has been quantified for Kenya, Uganda and Ethiopia.

3.3.4 Seed potato sources

Table 3.3 shows the sources of seed potatoes planted by farmers in Kenya, Uganda and Ethiopia. The dominant seed sources were the grower's own field and neighbours. In Uganda and Ethiopia the local market was an important source of seed potatoes, whereas this source was of less importance in Kenya.

In Kenya the number of farmers purchasing seed from seed growers was less than 1%, while in Uganda it was 4%. In West Shewa and Awi districts in Ethiopia it was comparably low with 3% and 2%, respectively. In North Shewa, however, potato farmers indicated they had planted 29% of potato fields with seed from a seed grower. Table 3.3 suggests a renewal rate with seed potatoes from outside the own farm of 30-70% per season. This may be an overestimation.

Table 3.3: Seed potato sources of farmers in Kenya, Uganda and Ethiopia (%)

	Kenya		Uganda		Ethiopia		
	Meru Central	Nyandarua	Kabale	Kisoro	West Shewa	North Shewa	Awi
Own field	30	70	61	40	54	32	65
Neighbour	66	28	16	15	12	10	3
Rural market	4	1	19	42	31	29	31
Seed grower	0	1	4	4	3	29	2
Number of respondents	235	498	247	101	157	129	398

3.3.5 Seed renewal period

Growers were asked whether they renewed seeds, and if so, after how many seasons. In Kenya, Uganda and Ethiopia 59%, 74% and 56% of the farmers indicated to never renew their seed potatoes respectively (Table 3.4). Those farmers who do renew their seed potatoes do so after an average 6, 7 and 3 seasons, from which it can be computed that only 7, 4 and 15% of the seed stock of Kenya, Uganda and Ethiopia respectively gets renewed each season from any of the possible sources outside the own farm.

Table 3.4: Percentage of farmers renewing their seed periodically, the average renewal period and the total fraction of potato seed renewed each season in Kenya, Uganda and Ethiopia.

	Kenya	Uganda	Ethiopia
Does not renew seeds periodically (%)	59	74	56
Does renew seeds periodically (%)	41	26	44
Average renewal period (seasons)	6	7	3
Seed stock renewed each season (%)	7	4	15

3.3.6 Seed potato management practices by ware potato producers

As the majority of seed potatoes planted originate from ware growers rather than from seed growers, it is worthwhile considering the seed potato management practices that are applied by ware potato growers.

Table 3.5 indicates that the majority of farmers in Kenya and Uganda select their seed from the bulk of potatoes at harvest. In Uganda, 17% of the farmers practice some form of in-field selection of the best plants for seed, and 11% produce their seed in a separate seed plot. In Ethiopia and Kenya, these technologies are also practised, but less frequently.

In Ethiopia, half of the potato farmers leave in the soil potatoes that will later be used as seed. This apparently is the method farmers prefer to store seed potatoes until the next season. Ethiopian farmers store their seed substantially longer than Kenyan and Ugandan farmers (Figure 3.1), as many only grow potatoes once a year. Kenyan potato farmers only store seed for 1–2 months before planting compared to 2–3 months by Ugandan potato farmers.

In Kenya and Ethiopia, 15 and 23% of the farmers indicated they store their seed under diffused light conditions, either in the house or in a special store whereas in Uganda 46% of farmers store seed potatoes under such conditions. Diffuse light assures seed potatoes with strong sprouts. In Kenya, 51% of the farmers store seed in a dark store, which they generally use for maize storage (Table 3.6).

Different techniques to stimulate the sprouting of seed potatoes, to assure they are ready on time for planting were mentioned (Table 3.7). In Kenya, burying potatoes in an underground pit is the most popular method. Covering the potatoes with crop residues, grass or manure is practised frequently in all three countries. Bringing the potatoes to a warm place in the house was used by more than 20% of the farmers in Ethiopia and Uganda, while also putting the seed potatoes in bags was frequently mentioned. Thirty two, 33 and 24% of farmers in respectively Kenya, Uganda and Ethiopia did not make any extra effort to stimulate sprouting.

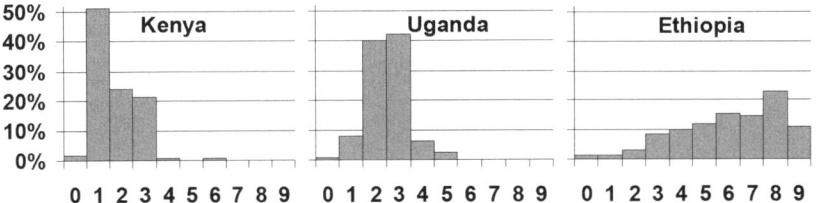

Figure 3.1: Storage duration (months) before replanting of seed potato harvested by ware potato growers in Kenya, Uganda and Ethiopia.

Table 3.5: Seed potato selection methods practiced in Kenya, Uganda and Ethiopia (% of farmers)

	Kenya	Uganda	Ethiopia
Separate seed plot	3	11	2
Positive selection[a]	3	17	9
Seed sized tubers from bulk harvest	92	72	40
Buy seeds from best plot neighbour	2	0	0
Preserve potatoes in the soil	0	0	49

[a] farmers selecting seeds from the best plants or best portion of their fields

Table 3.6: Seed potato storage techniques used by potato producers in Kenya, Uganda and Ethiopia (% of farmers)

	Kenya	Uganda	Ethiopia
Farmers storing seed potatoes	92.4	88.7	94.3
Storage method			
Diffuse light store	10	31	5
Dark store	51	16	2
In the field, not harvested	0	0	47
Dark space in the house	13	37	28
Light space in the house	5	15	18
Underground pit	21	1	0

Table 3.7: Dormancy breaking techniques used by farmers to force sprouting of seed potatoes in Kenya, Uganda and Ethiopia (% of farmers)

	Kenya	Uganda	Ethiopia
Wait	32	33	24
Put in an underground pit	29	5	0
Put in bags	7	18	29
Put in a warm place in the house	6	22	23
Cover with crop residues / grass / manure	26	16	17
Other	0	5	6

3.3.7 Specialized seed potato multiplication

In all three countries national research organizations are the sole source of modest amounts of disease free starter seed, meant for further multiplication by seed growers. In Kenya formal quality control and certification regulations exist for multiplication of starter seed, but few seed growers choose to make use of this system. They prefer to multiply seed potato without certification, and trade the seed to nearby farmers based on their reputation for quality seed potato. In Uganda a seed potato growers association with a self policing quality control system (Tindimubona et al., 2000) multiplies a significant proportion of the starter seed available, while the remainder is distributed through Non-Governmental Organizations (NGOs) to farmer groups, or is bought by individual seed potato multipliers. In Ethiopia starter seed is multiplied by farmer groups, collaborating with the potato research center (Getachew and Mela, 2000; German, 2006), without an elaborate quality control system.

3.3.8 Estimated volumes of seed potatoes in the specialized seed potato chain

A simple model was developed to calculate the final multiplication rate of the starter seed produced by the national research programmes (Figure 3.2) based on a few basic estimates. Average seed potato multiplication rate by seed growers was estimated fairly low at 7, considering the fact they generally do not have irrigation facilities. Not all starter seed ends up in the care of seed growers (Tindimubona et al., 2000). Also ware growers purchase starter seed, with the objective to obtain a new variety, or they receive it through an NGO or a development project. It was estimated that 60% of the starter seed is purposefully multiplied by specialized growers, while the other 40% is planted by ware potato farmers. Seed growers are further assumed to sell 60% of their seed yield after a single multiplication, 32% after the second multiplication, and 8% after a third multiplication.

Seed sold by multipliers is replanted several generations by ware potato farmers. Thiele (1999) found that in the Andes certified seed potato gave increased yields compared to local seed potatoes up to the third generation of re-use in farmer fields. Here we assumed a yield increase during only two seasons of re-use as the vast majority of high-quality seed in the three countries is not certified. Furthermore all seed was assumed to have lost its yield advantage compared to local seed after the fourth planting from being starter seed. As ware potato farmers sell their large tubers as ware potatoes, and only replant the smaller tubers, the average multiplication rate in ware potato farmers' fields was put at 2.

With these conservative estimates of the variables entered into the model (Figure 3.2), the total expected multiplication factor for high-quality starter seed injected into the seed system can be calculated as the sum of all grey squares. The grey squares represent high quality seed potatoes planted for ware potato production. In Table 3.8 the average amounts of starter seed produced in Kenya, Uganda and Ethiopia are used to calculate the total amount of high quality seed potatoes available yearly in the three countries. When projecting this against the calculated seed requirement in the three countries, this means that an estimated 1.6%, 2.4% and 1.3% of the total seed requirement is met by seed potatoes of relatively high quality, originating from basic seed initially sold by the national research organizations.

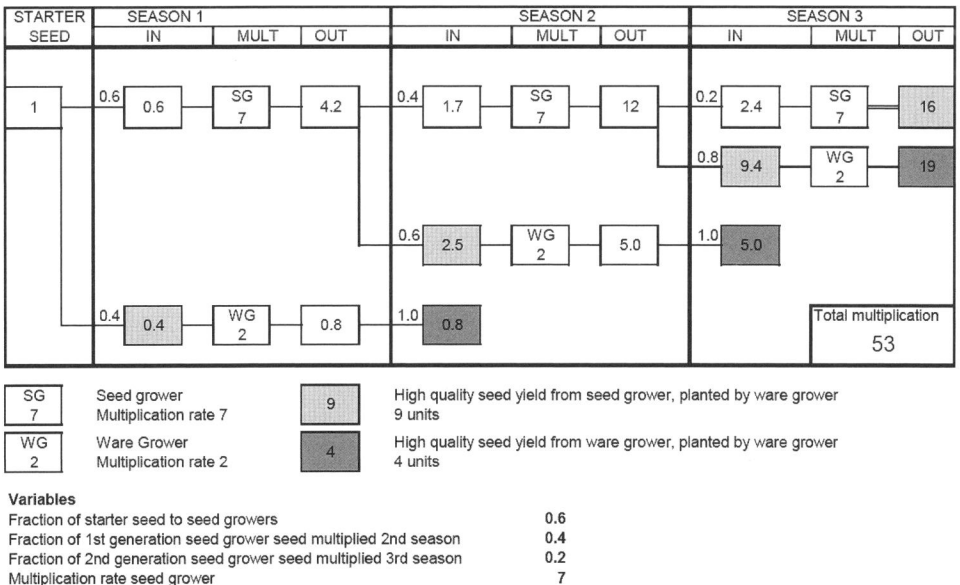

Figure 3.2: Model of starter seed potato multiplication rate in the seed potato systems of Kenya, Uganda and Ethiopia.

Table 3.8: Estimated amounts of high quality seed available to the potato growing systems in Kenya, Uganda and Ethiopia compared to the total seed potato requirement.

Country	Basic seed available for further multiplication (t/year)[a]	Estimated amount of high quality seed (t/year)[b]	Seed rate good quality seed (t/ha)	Area planted quality seed (ha/year)	Potato area (ha/year)	Proportion quality seed (%)
Kenya	71	3763	2	1882	116000	1.6
Uganda	103	5459	2.5	2184	90000	2.4
Ethiopia	78	4134	2	2067	161000	1.3

[a] Personal communication Mercy Wakahiu and John Karinga (Kenya); William Wagoire (Uganda) and Gebremedin Woldegiorgis (Ethiopia);
[b] Basic seed leaving the formal multiplication system is estimated to be multiplied a factor 53 before being degenerated to the level of "farmer seed" (Figure 4).

3.3.9 Seed potato economics

Table 3.9 presents economic data regarding seed potato use in Kenya, Uganda and Ethiopia. The acceptable price premium for high quality seed has been calculated on the basis of a marginal rate of return (MRR) to the farmer of 200%, meaning a farmer gets a net return of 2 dollars for every dollar he invests in seed potatoes. Although it has been suggested that a MRR of 100% can be considered sufficient for a farmer to decide to invest in a certain technology (Cassaday, 1988), we here prefer a higher minimum acceptable MRR because the return on the investment in high quality seed potato will not be achieved in a single season, but over a period of three seasons. For an urgently cash short smallholder potato farmer, a higher MRR is thus required as an incentive for technology adoption. Under this assumption price premiums on seed potatoes compared to the ware potato price of 26 to 87% are acceptable, under the current average yield levels and ware potato prices in the seven sample districts. On average, a price premium on high quality seed potato of 49% compared to price of ware potato, is acceptable.

Table 3.9: Acceptable price premium for high quality seed potatoes in Kenya, Uganda and Ethiopia

	Kenya		Uganda		Ethiopia			Average[c]
	Meru Central	Nyan-darua	Kabale	Kisoro	West Shewa	North Shewa	Awi	
Average yield (t/ha)	8.8	9.2	5.3	6.8	7.7	12.3	5.1	7.9
Production costs ($/ha)	510	475	278	244	366	358	394	375
Seed rate (t/ha)	1.5	1.6	1.1	1.0	1.1	1.1	1.6	1.3
Ware price ($/t)	80	45	88	72	62	79	91	74
Total yield gain (t/ha)[a]	2.1	2.2	1.2	1.6	1.8	2.9	1.2	1.9
Value yield gain ($/ha)	168	99	110	117	114	232	111	139
Min. MRR[b]	200	200	200	200	200	200	200	200
Acceptable investment ($/ha)	56	33	37	39	38	77	37	46
Acceptable seed price ($/t)	117	65	121	111	98	148	115	110
Acceptable additional cost seed	45	45	38	54	58	87	26	49
Additional cost seed over ware (%)	67	124	145	229	121	199	5	

[a] Estimated yield gain through use of high-quality seed potatoes over three seasons; generation 1, 2 and 3: 17%, 16% and 10% respectively (Thiele 1999); assumed area planted under improved seed 0.25, 0.5 and 1 ha in season 1, 2 and 3 respectively.
[b] Marginal Rate of Return; A MRR of 200% in a three-season period is considered fair to assure farmers are willing to risk investing in high quality seed potatoes.
[c] Average based on average yield, production cost, seed rate and ware price over the 7 sample districts.
[d] Current seed potato price as estimated by farmers (average of all sources of seed)

3.4 Discussion

3.4.1 Potato disease levels

The virus levels measured in the survey were substantially higher than those found in an earlier survey in Kenya (Were et al., 2003), in which less than 25% infection with PLRV was recorded. However, their figures were entirely based on visual observation, and samples were only taken from plants identified as infected, for confirmation through ELISA. This could have resulted in an underestimate of the actual infection rate. A survey in Uganda revealed PLRV and PVX levels of 17 and 35%, respectively (Kakuhenzire et al., 2000), which is also substantially lower than what was found in tubers from markets in Kenya. This study was, however, done from leaf samples rather than tuber samples, which could underestimate final tuber infection levels, especially when samples are taken earlier in the season. An extensive survey in Iran showed an average of 52% mixed infection rates in a similar leaf sample survey (Pourrahim et al., 2007). The results from the survey confirm observations of potato scientists in Kenya, that potato virus diseases are typically visible on between 20% and 80% of potato plants in ordinary potato fields (non-published field observations). The results strengthen the priority setting by potato farmers in Kenya, who put low seed potato quality at the top of the list of their problems (Gildemacher et al., 2006). It is likely that the high levels of mixed virus infection measured in the survey, are contributing to the low average potato yields in Kenya measured through the farmer yield estimates.

The data from the survey on bacterial wilt incidence and prevalence shows that the disease, considered a quarantine disease in many potato growing countries (Elphinstone, 2005), is endemic in the Kenyan potato farming system. Although the measured levels of incidence do not immediately seem alarming from an economic point of view, they can fluctuate dramatically according to the growing period of the crop in correlation with the relative humidity. In case of a severe outbreak of the disease affected farmers can consider their yield lost, as there is no chemical cure for infected plants. Farmers in Uganda and Kenya considered bacterial wilt a priority problem in their production (Gildemacher et al., 2009).

3.4.2 Seed potato management by ware producers

From the fraction of farmers that did renew seeds, in Kenya and Uganda the majority did so after eight seasons, compared to 3 seasons in Ethiopia. The shorter average renewal period in Ethiopia can be explained by the larger number of farmers growing potatoes only once a year, making seed potato storage more cumbersome (Eshetu et al., 2005) compared to Kenya and Uganda, where bi-modal potato farming is dominant.

The seed source data in Table 3.3 indicate a much higher seed renewal rate than Table 3.4. This is most likely related to the fact that many farmers do not make a distinction between seed source and variety source. As the process of seed degeneration and its underlying causes are poorly understood by smallholder potato growers (Crissman et al., 1993), producers may have mentioned the initial source of seed of the potato variety they were growing, rather than the place where their seed came from that very season. As a result the sources other than the own farm were overestimated. As such the seed renewal rates in Table 3.4 provide the most realistic estimate.

3.4.3 Seed potato economics

Potato producers in Kenya, Uganda and Ethiopia are, under the current circumstances in the seed and ware potato market, making an economically sound decision by not investing much in renewing their seed stock. In all districts accept Awi, the current cost of seed potatoes is higher than the calculated acceptable price for high quality seed potatoes (Table 3.9). However, these average figures are masking the fact that under specific circumstances, for example a higher ware potato price or higher yields through better production practices, regular investment in high quality seed potatoes can be economically feasible.

3.4.4 Improving seed quality in the local seed potato chain

As Thiele (1999) indicates, the local chain of seed potato production and marketing delivers an important service to potato producers. The local chain produced more than 95% of the volume of seed potatoes in the three countries. It has to be considered how the quality of seed potatoes from this source can be improved. This requires an alternative outlook on seed potato production in which each ware potato farmer is considered to contribute to the overall quality of seed potatoes in the system.

From the survey several opportunities for improving the management of seed potato quality by ware potato growers can be derived. Management of virus diseases and bacterial wilt by ware growers is a priority as the diseases are endemic in the potato production system. Seed potato storage technology used by ware potato producers can be improved. Furthermore better seed potato multiplication and selection techniques can be applied by ware growers. Multiplication of high quality seed potatoes in nurseries by ware potato producers using the 'small seed plot technique' (Aguirre et al., 1999; Kinyua et al., 2001; Kinyua et al., 2005), positive seed potato selection by ware potato farmers (Gildemacher et al., 2007) and diffused light storage (Potts, 1983) have shown to be useful technologies for improving seed potato quality management by ware growers.

Further research effort into the development of well adapted and marketable potato varieties with resistance to major potato viruses is highly relevant for the improvement of the quality of seed potatoes in the local chain. Further quantification of the importance of yield losses as a result of mixed infection with potato viruses is needed. This research shows very high incidences of four major yield reducing potato viruses in Kenya, but the actual yield loss as a result of these high levels of infection and the impact of resistance to some of these viruses deserves further study.

3.4.5 Improving efficiency in the specialized chain

In spite of the limited percentage of seed potatoes deriving from seed growers, the specialized chain should not be disregarded. Table 3.9 shows that quality seed potatoes can fetch an average 50% price premium under the current average production and prices and still provide the buyer with a healthy profit on his investment. Ware potato producers turning into informal seed potato multipliers for profit in Ethiopia (Eshetu et al., 2005; German, 2006), members of the Ugandan National Seed Potato Producers Association (UNSPPA) (Tindimubona et al., 2000), the short lived commercial success of large scale seed potato production by ADC in Kenya (Crissman et al., 1993), and individual Kenyan seed potato multipliers trading on their reputation, are the living proof of a potential for economically sustainable seed potato farming.

From Figure 3.2, entry points for increasing the impact of the specialized seed potato chain can be identified. The amount of starter seed produced and marketed can be increased, the fraction of starter seed multiplied at least twice by specialized seed potato multipliers can be improved, and also the multiplication rate of seed potatoes grown by seed growers could be higher.

Crissman et al. (1993) suggest commercialization of the production of starter seed as a strategy for increasing the effect of the specialized chain on the seed potato quality of the entire system. However, until there are other, commercial, actors taking on the role of starter seed potato producer, the small volumes produced by national research organizations are vital to the seed potato systems as the single source of tested disease free starter seed. Research organizations could improve the efficiency of their starter seed production by investigating new technological options for multiplication, well suited to the relatively small amounts currently required in the three countries. They can invest in the development of cost effective tissue culture techniques for rapid multiplication, cheap and easy-to-use disease testing methods and the adaptation of hydroponics and aeroponics techniques for minituber production.

Under the current circumstances the limited starter seed that is available is not multiplied with maximum efficiency (Tindimubona et al., 2000; Eshetu et al., 2005) largely because the number of skilled seed potato multipliers is at present small in all three countries. A larger number of commercial seed growers is desirable. The question remains as to how the further development of a class of seed growers can be facilitated, and how quality control systems can be shaped that answer to local needs.

All actors in this sector need to accept the presence of a local and a specialized seed potato chains. The local chain will continue to exist as it fits in a risk avoidance strategy by ware potato producers in reaction to uncertain ware potato prices, cash shortage and risk of crop failure. Simultaneously, the specialized seed potato chain requires further development to offer affordable high quality seed potatoes for those farmers who have the ability to realize its yield potential.

References

Aguirre, G., Calderon, J., Buitrago, D., Iriarte, V., Ramos, J., Blajos, J., Thiele, G., and Devaux, A. 1999. Rustic seedbeds: a bridge between formal and traditional potato seed systems in Bolivia. Impact on a changing world; International Potato Center Program Report 1997/ 1998. Lima: CIP

Cassaday, K. 1988. From agronomic data to farmer recommendations : an economics training manual. Mexico: CIMMYT.

Crissman, C.C., Crissman, L.M., and Carli, C. 1993. Seed potato systems in Kenya: a case study. Lima: CIP.

Elphinstone, J.G. 2005. The current bacterial wilt situation: A global overview. In: Allen, C., Prior, P., and Hayward, A.C. (eds). Bacterial wilt disease and the Ralstonia solanacearum species complex 9-28. St. Paul: APS press.

Eshetu, M., Ibrahim, O.E., and Etenesh, B. 2005. Improving potato seed tuber quality and producers' livelihoods in Hararghe, Eastern Ethiopia. Journal of New Seeds 7(3): 31-56.

FAO, 2008. http://faostat.fao.org/site/567/default.aspx, June 2008. Rome: FAO.

Fuglie, K.O. 2007. Priorities for potato research in developing countries: results of a survey. American Journal of Potato Research 84(5): 353-365.

German, L. 2006. Approaches for mountain regions: Operationalizing systems integration at farm and landscape scales. AHÍ working paper 21. Kampala: AHÍ.

Getachew, T. and Mela, A. 2000. The role of SHDI in potato seed production in Ethiopia: Experience from Alemaya integrated rural development project. In: 5th African Potato Association Conference, pp. 109-112. Kampala: APA.

Gildemacher, P., Demo, P., Kinyae, P., Nyongesa, M., and Mundia, P. 2007. Selecting the best plants to improve seed potato. LEISA Magazine 23(2): 10-11.

Gildemacher, P., Kaguongo, W., Ortiz, O., Tesfaye, A., Woldegiorgis, G., Wagoire, W., Kakuhenzire, R., Kinyae, P., Nyongesa, M., Struik, P., and Leeuwis, C. 2009. Improving potato production in Kenya, Uganda and Ethiopia: a system diagnosis. Potato Research, accepted.

Gildemacher, P.R., Maina, P., Nyongesa, M., Kinyae, P., Gebremedhin, W., Lema, Y., Damene, B., Shiferaw, T., Kakuhenzire, R., Kashaija, I., Musoke, C., Mudiope, J., Kahiu, I., and Ortiz, O. 2006. Participatory Analysis of the Potato Knowledge and Information System in Ethiopia, Kenya and Uganda. In: Sanginga, P.C., Bayer, A.W., Kaaria, S., Njuki, J., Wettasinha, C. (eds.). Innovation Africa: enriching farmers' livelihoods, pp. 203-219. London: Earthscan.

Kakuhenzire, R., Hakiza, J.J., Mateeka, B., Lemaga, B., Salazar, L., and Olanya, M. 2000. Incidence and importance of potato viruses in southwestern Uganda. In: 5th Triennial African Potato Association Conference, pp. 285-290. Kampala: APA.

Kinyua, Z.M., Olanya, M., Smith, J.J., El-Bedewy, R., Kihara, S.N., Kakuhenzire, R.K., Crissman, C., and Lemaga, B. 2005. Seed-plot technique: empowerment of farmers in production of bacterial wilt-free seed potato in Kenya and Uganda. In: Bacterial wilt disease and the Ralstonia solanacearum species complex.

Kinyua, Z.M., Smith, J.J., Lung'aho, C., Olanya, M., and Priou, S. 2001. On-farm successes and challenges of producing bacterial wilt-free tubers in seed plots in Kenya. African Crop Science Journal 9(1): 279-285.

Monares, A. 1987. Analytical framework for design and assessment of potato seed programs in developing countries. The social sciences at CIP Report of the third social science planning conference, held at CIP, Lima, Peru, September 7 10, 1987.

Potts, M.J. 1983. Diffuse light potato seed storage as an example of technology-transfer - a case study. American Potato Journal 60(4): 217-226.

Potts, M.J. and Nikura, S. 1987. Seed potato farm developed in Burundi encourages new ideas and techniques. Agriculture International 39(9-10): 266-270.

Pourrahim, R., Farzadfar, S., Golnaraghi, A.R., and Ahoonmanesh, A. 2007. Incidence and distribution of important viral pathogens in some Iranian potato fields. Plant Disease 91(5): 609-615.

Reestman, A.J. 1970. Importance of the degree of virus infection for the production of ware potatoes. Potato Research 13(4): 248-268.

Salazar, L. and Jayasinghe, U.E. 2002. Techniques in plant virology at CIP. Lima: CIP.

Struik, P.C. and Wiersema, S.G. 1999. Seed potato technology. Wageningen: Wageningen University Press.

Thiele, G. 1999. Informal potato seed systems in the Andes: why are they important and what should we do with them? World Development 27(1): 83-99.

Tindimubona, S., Kakuhenzire, R., Hakiza, J.J., Wagoire, W.W., and Beinamaryo, J. 2000. Informal production and dissemination of quality seed potato in Uganda. . In 5th African Potato Association Conference, pp. 99-104. Kampala, APA.

Turkensteen, L.J. 1987. Survey of diseases and pests in Africa: fungal and bacterial diseases. Acta Horticulturae (213): 151-159.

Were, H.K., Narla, R.D., Nderitu, J.H., and Weidemann, H.L. 2003. The status of potato leafroll virus in Kenya. Journal of Plant Pathology 85(3): 153-156.

4 Seed potato quality improvement through positive selection by smallholder farmers in Kenya

P.R. Gildemacher[a,b], E. Schulte-Geldermann[a], D. Borus[a], P. Demo[a], P. Kinyae[c], P. Mundia[d], P.C. Struik[e]

[a] International Potato Center (CIP), Nairobi, Kenya
[b] Royal Tropical Institute (KIT), Amsterdam, The Netherlands
[c] Kenya Agricultural Research Institute (KARI), Tigoni, Kenya
[d] Jomo Kenyatta University of Agriculture and Technology (JKUAT), Thika, Kenya
[e] Centre for Crop Systems Analysis, Department of Plant Sciences, Wageningen University and Research Centre (WUR), Wageningen, The Netherlands

Published in: Potato Research (2011) 54:253-266

Abstract

In Kenya, seed potato quality is often a major yield constraint in potato production as smallholder farmers use farm-saved seed without proper management of seed-borne pests and diseases. Farm-saved seed is therefore often highly degenerated. We carried out on-farm research to assess whether farmer-managed positive seed selection could improve yield. Positive selection gave an average yield increase in farmer managed trials of 34%, corresponding to a 284 Euro increase in profit per hectare at an additional production cost of only 6 Euro per ha. Positive selection can be an important alternative and complementary technology to regular seed replacement, especially in the context of imperfect rural economies characterized by high risks of production and insecure markets. It does not require cash investments and is thus accessible for all potato producers. It can also be applied where access to high quality seed is not guaranteed. The technology is also suitable for landraces and not recognized cultivars that can not be multiplied formally. Finally the technology fits seamlessly within the seed systems of sub-Sahara Africa, which are dominated by self-supply and neighbour supply of seed potatoes.

4.1 Introduction

4.1.1 Seed potato systems in Kenya

Potato (*Solanum tuberosum* L.) is the second most important food crop in Kenya in terms of bulk harvested; it is an important staple and cash crop for smallholder farmers in the Kenyan highlands. Poor seed potato quality is a major yield reducing factor in potato production in Kenya (Gildemacher et al. 2009a,b). In a survey by Fuglie (2007) viruses and bacterial wilt caused by *Ralstonia solanacearum* scored as important priorities for action in the eyes of potato researchers, whereas nematodes scored much lower. Improving seed potato quality is considered a pathway to improve smallholder potato yields and income (Getachew and Mela 2000; Tindimubona et al. 2000; Eshetu et al. 2005; Hirpa et al. 2010).

Seed potato health is a major determinant of the yield potential of a potato crop. Over generations seed potato quality degenerates as a result of tuber-borne diseases, among which viruses play an important role (Salazar 1996). Turkensteen (1987) identified bacterial wilt, caused by *R. solanacearum* and the viruses PVY and PLRV as seed borne potato diseases of major importance in Central Africa (Rwanda, Burundi and Eastern DRC), but also mentioned soft rot (*Pectobacterium chrysanthemi*), Fusarium wilt and dry rot (*Fusarium solani*) and Verticilium wilt (*Verticilium albo-atrum*) as being of economic importance. Solomon-Blackburn and Barker (2001) mention PVY and PLRV as most important viruses worldwide, and PVX as relatively mild as single infection, but potentially damaging in combination with other viruses.

Degeneration over seed generations is the combined result of increasing percentage of seed tubers infected, increasing number of tubers infected with multiple viruses and an increasing concentration of particles of these viruses in the seed tubers.

The speed with which the yield potential of the seed stock degenerates over generations of re-use as a result of accumulation of viruses depends on a number of factors. Firstly, the disease pressure is related to the abundance of the vector of viruses, most often aphids, and the number of diseased plants present. At high temperatures viruses reproduce faster within the plants and most virus vectors also have a shorter generation duration and are also more active than at lower temperatures, thus increasing the disease pressure. Secondly the degeneration depends on the variety grown. Varieties differ in levels of resistance to virus infection and virus particle multiplication within the plant (Salazar 1996). Moreover, some varieties tolerate virus infection better than others, which is reflected in lower yield losses under similar virus incidences.

Yield loss can be avoided through regular replenishment of seed stocks by high-quality seed potatoes multiplied by specialist growers from disease–free starter material. The specialized production skills, distribution system and quality control system required, combined with the low multiplication rate, the bulkiness and the poor shelf life of seed potatoes, all make high-quality seed potatoes expensive. Seed potatoes represent a major component of potato production costs.

In Northern countries, where producers have fairly reliable market outlets and relatively predictable yields, the return on investments of high-quality seed potatoes by ware potato producers is positive. In Sub-Sahara Africa however, yields and profits fluctuate widely as a result of variation in rainfall patterns and unreliable market chains. Investment in planting high-quality seed potatoes is therefore less attractive for Sub-Sahara African farmers. Furthermore smallholder farmers lack the cash required for investment in high-quality seed potatoes. Rather than relying on specialized seed potato growers, the seed potato systems in Sub-Sahara Africa are dominated by neighbour and self-supply (Crissman et al. 1993; Gildemacher et al. 2009b; Hirpa et al. 2010).

Seed system interventions to improve smallholder potato yields have been initiated in many developing countries. Some interventions introduced and supported formal certified seed potato production schemes with independent quality control like in Kenya (Crissman et al. 1993), Rwanda, Bolivia and Peru, while others focussed on building informal, non-certified, farmer-based seed potato multiplication schemes, like for example in Uganda (Tindimubona et al. 2000). Invariably, interventions were based on a model of specialised seed potato growers as suppliers of high-quality seed potatoes to smallholder ware potato farmers. This could be considered as attempts to transfer successes of specialized seed potato multiplication systems in Northern countries, like the Netherlands, UK and Canada (Young 1990), to developing countries. Notwithstanding pilot successes with building such seed potato multiplication systems in developing countries, there is little evidence of cases where building a specialized seed potato system has led to drastic and sustainable improvement of the yields of poor potato producers.

During the temporary successful operation of large scale seed potato multiplication and distribution in Kenya between 1980 and 1990, this only accounted for about 1% of all seed potatoes planted in the country (Crissman et al. 1993). Excluding the well established seed potato industry in South Africa, there are currently no examples of Sub Saharan African countries, with the agro-ecology suitable for seed potato production, that satisfy a substantial proportion of their demand for seed potatoes through formal certified or otherwise quality controlled seed production. Gildemacher et al. (2009b) calculated that in Kenya, Uganda and Ethiopia the proportion of seed potatoes originating directly or indirectly from quality controlled multiplication was less than 3% of the total seed requirement. In spite of the undisputed importance of high quality seed potatoes as an input for intensive potato production, it is apparently difficult to make commercial high quality seed potatoes available to the majority of potato producers.

Considering the importance of farm saved seed potatoes in Eastern Africa, Gildemacher et al. (2009b) identified the need to improve seed potato quality management by ware potato producers as a component of improving the overall quality of seed potatoes used. This raised the following question: what technologies can smallholder potato farmers apply to maintain or even improve the quality of their own seed potato stocks?

4.1.2 Positive seed potato selection

Positive selection is an old technology that was used primarily in formal seed potato multiplication to select mother plants from the best plot of potatoes as the starting point of the multiplication system (De Bokx and Van de Want 1987). The best potato plants in a field are marked before crop senescence that obscures disease symptoms. The marked plants serve as mother plants for seed potatoes used for the next season's potato crop. Positive selection has been used in Central Africa as the starting point for a seed multiplication system (Haverkort 1986). Positive selection is now widely regarded as an obsolete technology in formal seed potato production systems. Currently seed potatoes in formal seed systems are multiplied from tested, disease–free, tissue-culture material or from other nuclear stock which has been proven to be disease free. The use of positive selection as an on-farm method to maintain seed potato quality is also mentioned in literature (Struik and Wiersema 1999), but is not commonly used by ware potato producers, nor is its use promoted.

A specific action research programme on positive seed potato selection was implemented by the International Potato Center (CIP) from 2004 till today. The main focus of the programme was the training of ware potato farmers in positive seed potato selection (Gildemacher et al. 2007). The positive selection initiative integrated research and development objectives into a single effort, aiming at innovation rather than research results alone.

This paper presents the results of farmer managed trials in which positive selection is compared to common farmer practice. It demonstrates that the technology can provide an additional option for smallholder potato producers to manage the quality of their seed potatoes. The paper goes on to discuss the likely causes of the observed increases in production.

4.2 Materials and Methods

The positive selection technology was tested under full farmer control, minimizing the scientist influence on trial execution. It was not the technological soundness of positive selection that had to be proven, but rather the value of the technology in the hands of smallholder potato producers in the Kenyan production system. The technology had been in existence for decades, but was never adopted on a large scale by ware potato growers. Surmising that this could not be the result of the complexity nor the lack of efficiency of the technology, but rather the lack of effective training and promotion, a great deal of attention was put on the development of a training methodology that could potentially be scaled out to a national level.

4.2.1 Training methodology

A training approach for farmer groups, resembling farmer field schools (FFSs), was used for the training in positive selection, with some deliberate differences. The positive selection training was less intensive than usual in FFSs to minimize the required facilitator and farmer time. The meetings of the farmer groups were more facilitator-led than usual in FFSs and the agenda was fixed by the programme. Rather than trying to cover a diversity of potato issues a specific choice was made on seed potato quality management and seed-borne diseases. The training of the facilitators was limited to 2 days, with further support and interaction on-the-job. The total number of meetings of the farmer group was 9 times over a period of roughly 10 months. The demonstration experiment comparing farmer practice with positive selection formed the centre of the farmer group training.

It was attempted to meet both the scientist and the farmer need for experimentation simultaneously. The set-up of the trials was such that it resembled most the manner in which a farmer would experiment without involvement of scientists (Bentley 1994). Positive selection was tested, against the current farmer practice (see Box 4.1). Replications were over farmer groups rather than within farmer groups.

Seed potatoes were obtained by the farmer group from an existing potato field of at least 1,000 m^2, planted with a popular variety, and considered to be representative of their potato fields. The field was divided into two equal portions. One half was designated to source seed potatoes using positive selection, the other half using farmer common practice. For positive selection the farmers pegged the best looking plants as they were taught, just before flowering, roughly 10 weeks after planting. Two weeks later the farmers inspected the field and removed pegs from plants with newly developed disease symptoms. Pegged plants were harvested individually and plants with few, small or misshaped tubers were rejected. Tubers of 25-90 mm from the remaining pegged plants were collected as seed potatoes for the positive selection treatment of the demonstration trial. Seed potatoes for the farmer practice treatment were selected from the bulk of potatoes harvested from the other half of the field, according to common farmer practice. Seeds from both sources were stored next to each other using the common farmer practice.

> **Box 4.1 Setup of the farmer-managed positive selection experiments**
>
> Positive selection trail set-up
> 1. Let the group of farmers select an average potato field.
> 2. Divide it into two and let farmers peg healthy looking plants just before flowering in half of the field; reconfirm the health status of pegged plants two weeks later.
> 3. Harvest seed potatoes after judging the tubers of each pegged plant in the positive selection plot; select seed from the farmer practice plot using common farmer practice.
> 4. Store seed potatoes from both sources under the same conditions.
> 5. Plant an equal number of the positive selection and farmer selection seeds in adjacent plots, perpendicular to the slope.
> 6. Monitor the experiment; let the group of farmers practice positive selection once more.
> 7. Harvest separately the two plots, record total weight and evaluate.
>
>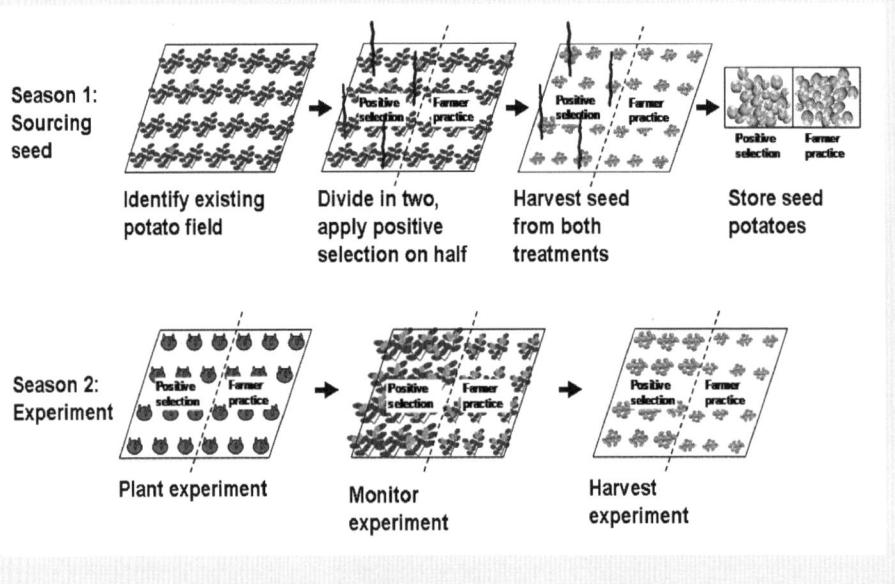

For the experiment a field supplied by the farmer group was divided into two, perpendicular to the slope. One half was planted with seed tubers obtained through positive selection, the other with farmer practice derived seed tubers. Planting, fertilizer application, disease control, hilling and weeding were all done by the farmer groups using their common practice.

4.2.2 Data collection and analysis

Eight weeks after planting a random sample of minimum 400 plants was inspected for visual virus and bacterial wilt symptoms. The number of plants showing symptoms as well as the total number of plants was recorded. At harvest the total number of plants in both plots was counted, and both plots were harvested. Marketable yield of the plots was recorded as all tubers above 25 mm.

Here data are presented from two separate seasons of farmer experimentation, the short rainy season of September 2005-February 2006 and the long rainy season of April-August 2010.

In 2005-2006 yield data could be collected from 13 farmer groups and 12 trials yielded useful disease data. Reasons for rejecting trials were several, including incorrect data collection by the teams of farmers and extension staff, harvest of the trial by thieves, destruction of the field by porcupines, a differential treatment during the growing season or separate seed storage of one of the two treatments and complete crop failure as a result of drought.

For 2010 yield data was available from 72 trials, but only in a selection also disease data were collected. A first selection of trials for data analysis was made by taking those trials having both yield and disease data. Furthermore trials that did not yield more than 3 t/ha for the farmer selection treatment were omitted from the analysis of yield differences. Trials from Eldoret East were omitted from the analysis of disease data because of irregularities in field data collection.

Paired sample t-tests were conducted to evaluate the effect of positive selection compared to farmer selection on yield as well as farmer-scored visual bacterial wilt and virus incidence.

4.3 Results

4.3.1 Experimental results

In 2005-2006 the positive selection plots gave an average yield of 14.2 t ha^{-1} which was significantly higher than the 11.8 t ha^{-1} for the farmer seed selection plots. The average yield increase of positive selection over farmer selection was 28% (Table 4.1). In 2010 the positive selection lots gave on average 13.1 t ha^{-1}, compared to 8.6 t ha^{-1} for the farmer seed selection, a yield increase of 53% (Table 4.2).

Figure 4.1 shows that the effect of positive selection was apparent notwithstanding the yield of the farmer selection treatment. The average yield increase for the 25% lowest yielding trials was 2.7 t/ha, which is a 55% increase. The average yield increase for the 25% highest yields was 5.1 t/ha, which represents a 29% yield increase.

Table 4.1: Yield of positive seed selection plots and farmer selection plots in farmer managed trials in Nyandarua and Nakuru districts, Kenya, October 2005 to February 2006.

Farmer Group	Yield (t/ha)		Yield increase due to positive selection (%)
	Positive selection	Farmer selection	
Dundori	21.4	21.3	0
Elburgon	16.3	8.9	84
Gilgil (Eburru)	19.2	10.1	91
Gitiri (North Kinangop)	11.1	8.3	35
Heni	20.2	21.0	-4
Kipipiri	11.9	9.0	32
Kirima	8.7	7.6	13
Kuresoi	12.5	11.8	6
Munyaka (Bahati)	16.8	13.6	24
Njoro	11.7	10.4	13
Ol Kalou 2	4.0	3.9	4
Olenguruone	23.4	22.6	4
Subukia (Mbogoini)	7.5	4.7	58
Average	**14.2**	**11.8**	**28**
t-value		3.02	
Degrees of freedom		12	
P		0.005	

Table 4.2: Yield of positive seed selection plots and farmer selection plots in farmer managed trials in Eldoret East, Kiambu East, Mt. Elgon and Transmara districts, Kenya, May 2010 to September 2010.

Farmer Group	Yield (t/ha)		Yield increase due to positive selection (%)
	Positive selection	Farmer selection	
Eldoret East 1	8.0	4.8	67
Eldoret East 2	7.5	6.0	25
Eldoret East 3	8.7	6.0	44
Eldoret East 4	8.0	6.7	20
Eldoret East 5	13.0	7.9	65
Kiambu East 1	12.8	5.0	156
Kiambu East 2	6.0	5.0	20
Kiambu East 3	8.8	6.8	29
Kiambu East 4	11.2	7.2	56
Kiambu East 5	8.0	7.2	11
Kiambu East 6	12.6	8.0	58
Kiambu East 7	15.0	8.0	88
Kiambu East 8	11.0	9.2	20
Kiambu East 9	13.2	9.2	43
Kiambu East 10	18.2	11.8	54
Kiambu East 11	20.0	17.6	14
Mt Elgon	6.5	4.0	63
Transmara 1	23.6	15.6	51
Transmara 2	37.3	16.9	121
Average	**13.1**	**8.6**	**53**
t-value		4.33	
Degrees of freedom		18	
P		<0.001	

Table 4.3: Farmer scored virus and bacterial wilt infection rates of positive seed selection plots and farmer seed selection plots in farmer managed trials in Nyandarua and Nakuru districts, Kenya, October 2005 to February 2006.

Farmer group	Farmer virus incidence scores (%)		Incidence of wilted plants (%)	
	Positive selection	Farmer selection	Positive selection	Farmer selection
Elburgon	5.4	9.1	0.3	6.3
Gilgil	4.7	7.9	0.0	0.3
Kirima	4.6	8.8	0.0	0.0
Kuresoi	3.6	6.8	8.2	16.9
Molo	4.3	11.1	1.2	4.1
Gitiri	5.4	9.8	0.0	0.0
Njoro	5.1	14.9	0.8	1.1
Ol Kalou 2	5.0	9.1	0.0	0.0
Olenguruone	2.0	3.6	0.1	1.1
Pagma Naivasha	8.1	8.8	0.0	0.0
Rurii Ol Kalou	7.3	9.7	0.4	2.5
Shamba Ndogo	5.0	9.0	0.4	0.0
Average[a]	5.0	9.0	1.3	3.5
t-value		5.87		2.32
Degrees of freedom		11		7
P		<0.001		0.027

[a] Average incidence and t-value of bacterial wilt infection calculated as a function of those fields that did have a detected infection

Table 4.3 and Table 4.4 show that the visible virus incidence as scored by the farmers in the demonstration trials was significantly reduced as a result of positive selection, from 9% to 5% in 2005-06 and from 18.8% to 7.1% in 2010.

Also bacterial wilt incidence was significantly lower in the positive selection plots than in the farmer selection plots (Table 4.3 and Table 4.4). In those trials in 2005-06 where bacterial wilt was observed, the positive selection plots had an average incidence of 1.3% compared with 3.5% in the farmer selection plots. In 2010 positive selection reduced bacterial wilt infection from an average of 7.6% to 2.6%.

Table 4.4: Farmer scored virus and bacterial wilt infection rates of positive seed selection plots and farmer seed selection plots in Kiambu East, Mt. Elgon and Transmara districts, Kenya, May 2010 to September 2010.

Farmer group	Farmer virus incidence scores (%)		Incidence of wilted plants (%)	
	Positive selection	Farmer selection	Positive selection	Farmer selection
Kiambu East 1			0	2
Kiambu East 2			1	2
Kiambu East 3			0	1
Kiambu East 4			0	3
Kiambu East 5			2	2
Kiambu East 6			0	2
Kiambu East 7			0	1
Kiambu East 8			0	1
Kiambu East 9			0	3
Kiambu East 10			0	1
Kiambu East 11			0	2
Mt Elgon 1	2	9	5	18
Mt Elgon 2	3	8	7	11
Mt Elgon 3	6	11	8	19
Mt Elgon 4	6	16	7	22
Mt Elgon 5	7	26	6	21
Mt Elgon 6	27	34	14	23
Transmara 1	3	21	0	8
Transmara 2	3	25	0	2
Average[a]	7.1	18.8	2.6	7.6
t-value		4.79		4.15
Degrees of freedom		7		18
P		0.001		0.0003

[a] Average incidence and t-value of virus infection calculated as function of those fields with observations. Fields without figures represent trials in which no virus incidence levels were scored.

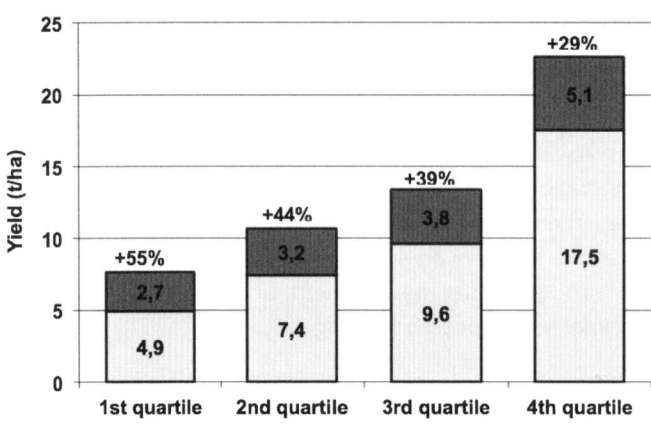

Figure 4.1: Absolute and relative average yield increase as a result of positive selection compared to farmer selection for the demonstration trials of 2005-06 (n=13) and 2010 (n=18) combined, per quartile of farmer selection yield level.

4.3.2 Economic analysis

Table 4.5 shows that the average yield increase obtained in the trials was 3.5 t/ ha. This yield increase gave an increase in the gross benefit per hectare of Euro 290. The required investment in additional labour was estimated at 4 man days per hectare, costing 6 Euro. Adopting the technology would result in an estimated marginal net benefit of Euro 284 per hectare at a moderate farm gate price level of 900 Ksh / 110 kg bag of potatoes. The return on labour, provided the smallholder farmer would invest his or her own time in positive selection, is 70.9 Euro per day of labour, roughly 46 times the cost of labour.

Table 4.5: Marginal net benefit of the adoption of positive selection based on the average yields obtained in 2005-06 and 2010.

	Positive selection	Farmer practice
Price per kilo ( )[a,b]	0.08	0.08
Productivity (t/ha)[c]	13.7	10.2
Gross benefit (/ha)	1,142	853
Additional labour cost (/ha)[d]	6	-
Marginal net benefit (/ha)	**284**	-
Return on labour (/day)	**70.9**	

[a] Estimated minimum farm gate price (900 Ksh / 110 kg bag).
[b] 1€=97.79 Ksh at www.oanda.com, 01/09/2010.
[c] Average over the 2005-06 and the 2010 season trial data
[d] Casual labour estimated at 150 Ksh / day.

4.4 Discussion

Positive selection showed to be a valuable technology for smallholder producers. Potato yields in the demonstration trials under full farmer management were significantly increased as a result of the use of positive seed potato selection by producers. Based on the results from the two seasons of demonstration trials presented here, a yield increase under Kenyan conditions between 28% (2005-06) and 54% (2010) can be expected.

These yield increases have been obtained over a wide variety of circumstances, such as different varieties and locations. As such, the technology has proven to be robust and effective notwithstanding the variation in circumstances. The technology substantially increased yields in situations where the farmer selection yielded very poorly, but also there where yield were already well above the Kenyan average of about 9 t/ha. In addition, the yield increase was obtained through farmer management, under circumstances very much representative of Kenyan smallholder potato farming.

Most importantly, this yield increase could be obtained without any additional cash investment, which is of essential importance for cash-poor smallholder farmers. The additional labour required for cutting pegs, pegging the healthy plants and harvesting the pegged plants one by one yields an estimated return of 70.9 Euro per man-day, which is 46 times the estimated cost a casual labour in the Kenyan countryside.

From a scientific point of view the question remains why positive selection has worked. Considering the data collected by the farmers themselves with respect to virus and bacterial wilt incidences, positive selection can reduce both significantly. A substantial reduction in virus and bacterial wilt infection could well be a contributing factor to the yield increases observed. The reduction of seed-borne pathogens other than virus and bacterial wilt disease may have contributed to the yield difference observed between positive selection and farmer practice. Turkensteen (1987) identified *Erwinia* spp. bacteria (nowadays called *Pectobacterium* spp.) and *Fusarium* spp. fungi as 'important' seed-borne pathogens in Central Africa.

Virus incidence levels were scored by producers, after a very basic explanation of virus symptoms by an extension worker. The observed virus infection levels by the farmers could well be much lower than the real infection levels. Visual virus infection detection is not all that easy for experts, let alone for potato farmers who have been briefly trained in the field. Measurements of virus incidences are scarce, but in a quick survey of seed potatoes sold in rural markets in Kenya an average incidence of 71, 75, 57 and 41% for PLRV, PVY, PVX and PVA respectively was recorded (Gildemacher et al, 2007b). This is substantially higher than the incidence scores by the producers presented in this paper.

During the training of farmers it quickly became apparent that the identification of diseased plants requires thorough understanding and experience, which is not easy to obtain in a few group trainings. Fortunately positive selection is based on the identification of the most healthy plants, or 'select the best' (Gildemacher et al., 2007a). Selecting the best looking plants in a potato field is far easier than identifying accurately each and every diseased plant.

Another possible contributing factor to the increase in yield through the positive selection treatment may be related to an unintentional shift in seed tuber size. As only a selection of plants gets pegged by the producers, they are under pressure to accept both smaller as well as larger tubers as seed potatoes than they are normally inclined to plant. Average seed tuber weight was not recorded specifically in the farmer run trials. The facilitators instructed producers to not go for tubers smaller than 25 mm, which is smaller than the average tuber size planted in Kenya, and not beyond 90 mm, which is larger than the average tuber size. In the farmer-managed plots producers were selecting according to common farmer practice, which is to select 'egg sized' tubers from the bulk of potatoes harvested. It cannot be ruled out that part of the yield effect is caused by an unintentional shift in tuber size, and not by reducing seed borne diseases alone.

The increases in yields are the result of a single season of positive selection, by farmers with no prior experience in practicing this technology. It may be expected that yield differences could increase if further positive selection was practiced consistently over several seed generations. A first indication of a potential of add-on effects over seasons is that positive selection assured a 55% average yield increase for the 25% lowest yielding trials, but still a 29% increase for the 25% highest yielding plots.

There is a large variation in the yield increase obtained through the technology. This can partly be attributed to differences in effectiveness of training of the farmer groups which may have different causes, ranging from the motivation and capacity of the facilitator of the public extension staff assisting the group, to the motivation and cohesiveness of the farmer groups involved. Other causes for variation may include the different disease incidence and varieties grown in the fields where the seed potato was sourced for the experiment.

The farmer managed trials with positive selection convincingly show that the technology can substantially increase smallholder potato productivity. This does however not automatically mean that it is the best possible solution for potato farmers. The regular replacement of a farmers' seed stock with high quality seed potatoes from a more formal seed multiplication system may well be more economic. A number of demonstration trials in which an additional plot was planted with certified seed (data not presented) indicate that this can outyield positive selection under most circumstances. For farmers who can afford the risk of investing in certified seed potatoes because they have yield security, the required cash and a fairly sure market, buying high quality seed regularly is probably more economic than practicing positive selection. This does assume that these farmers have access to these certified seeds of the right variety at the right time.

Considering the simplicity of positive selection and the apparent good fit within the prevailing informal seed potato system of Kenya, one has to question why potato farmers have not been practicing this technology all along. A number of reasons can be identified. An important first reason is the limited understanding of seed degeneration and the role of potato viruses in this among farmers and extension staff alike. The training on positive selection and viruses was an eye-opener that made farmers and extension staff aware of the poor health status of the average Kenyan potato field. The knowledge tests implemented before the training of both trainers and farmers in 2005-06 confirmed this limited understanding (data not shown). Secondly the potato crop dies off before harvest and its product is found below ground. In maize production positive selection of the best cobs for next season's planting is well known and widely practiced for open pollinated varieties. For potato this is more complicated, as plants need to be pegged before senescence sets in, and ideally before the crop closes and starts flowering. Finally positive selection was not seen as a technology suitable for large scale seed potato multiplication, with good reason. For a specialized seed multiplier positive selection can only be of assistance in selecting mother plants in the first generations. In later generations the removal of sick plants from a largely healthy field (negative selection) is the only possible way to bulk-up seed of high quality for commercialization.

Seed potato quality management has invariably been addressed through specialized seed multipliers. This has provided an excellent solution to seed potato degeneration in Northern countries. In less perfectly functioning rural economies of developing countries, with higher production risks and market insecurities, specialised multiplication systems have been much less successful. Decision makers in seed potato programs and projects have been focussing fully on seed *multiplication*, and not paid enough attention to the potential of seed potato quality *maintenance by non-specialized ware potato farmers*.

A number of research question with regard to the mechanisms behind the success of the positive selection technology remain. In addition it would be of interest to consistently continue positive selection over a number of seasons in the same potato plant population to assess the potential of the technology to further increase the yield potential over several generations. To study the technology in further detail more thorough replicated trials under controlled conditions, quantifying the virus load in the plant population would be recommended. Furthermore it would be advisable to control more rigidly the stability of seed tuber size in these replicated trials.

Nevertheless the results of the farmer managed trials demonstrate that positive selection is an important alternative and complementary technology to regular seed replacement. In the first place it does not require cash investments and is thus accessible for all potato producers. Secondly it can be applied where access to high quality seed is not guaranteed. Thirdly the technology is also suitable for landraces and cultivars that are not officially recognized and can thus not be multiplied formally. Fourthly the technology fits seamlessly within the currently most important seed system of sub-Sahara Africa, which is dominated by self-supply and neighbour supply of seed potatoes.

References

Crissman CC, Crissman LM, Carli C (1993) Seed potato systems in Kenya: a case study. CIP, Lima

De Bokx J, Van de Want J (1987) Viruses of potatoes and seed-potato production. Pudoc, Wageningen

Eshetu M, Ibrahim OE, Etenesh B (2005) Improving potato seed tuber quality and producers' livelihoods in Hararghe, Eastern Ethiopia. J New Seeds 7(3):31-56

Fuglie, K.O. 2007. Priorities for potato research in developing countries: results of a survey. Am J Potato Res 84:353-365.

Getachew T, Mela A (2000) The role of SHDI in potato seed production in Ethiopia: Experience from Alemaya integrated rural development project. In: Adipala E, Nampala P, Osiru, M (ed) Proceedings of the 5th Triennial Congress of the African Potato Association, May 29-June 2, 2000, Kampala, Uganda, pp 415-419

Gildemacher P, Demo P, Kinyae P, Wakahiu M, Nyongesa,M, Zschocke T (2007a) Select the best: positive selection to improve farm saved seed potatoes; Trainers manual. International Potato Center, Nairobi.

Gildemacher P, Mwangi J, Demo P, Barker I.(2007b) Prevalence of potato viruses in Kenya and consequences for seed potato system research and development. In: Khalf-Allah, AA et al (eds). Proceedings of the 7th Triennial Congress of the African Potato Association, October 22-26, 2007, Alexandria, Egypt, pp. 238-241

Gildemacher P, Kaguongo W, Ortiz O, Tesfaye A, Woldegiorgis G, Wagoire W, Kakuhenzire R, Kinyae P, Nyongesa M, Struik PC, Leeuwis C (2009a) Improving potato production in Kenya, Uganda and Ethiopia: a system diagnosis. Potato Res 52 (2): 173-205

Gildemacher PR, Kaguongo W, Demo P, Barker I, Wagoire WW, Wakahiu M, Woldegiorgis G, Leeuwis C, Struik PC (2009b) A description of seed potato systems in Kenya, Uganda and Ethiopia. Am J Potato Res 86: 373-382

Haverkort A (1986) Forecasting national production improvement with the aid of a simulation model after the introduction of a seed potato production system in central Africa. Potato Res 29(1): 119-130

Hirpa A, Meuwissen MPM, Tesfaye A, Lommen WJM, Oude Lansink A, Tsegaye A, Struik PC (2010) Analysis of seed potato systems in Ethiopia. Am J Potato Res 87:537-552

Salazar LF, (1996) *Potato viruses and their control*. International Potato Center, Lima.

Solomon-Blackburn RM, Barker H. (2001) Breeding virus resistant potatoes (*Solanum tuberosum*): A review of traditional and molecular approaches. Heredity 86:17-35

Struik PC, Wiersema SG (1999) Seed potato technology. Wageningen University Press, Wageningen:.

Tindimubona S, Kakuhenzire R, Hakiza JJ, Wagoire WW, Beinamaryo J (2000) Informal production and dissemination of quality seed potato in Uganda. In: Adipala E, Nampala P, Osiru, M (ed) Proceedings of the 5th Triennial Congress of the African Potato Association, May 29-June 2, 2000, Kampala, Uganda, pp 99-104

Turkensteen LJ (1987) Survey of diseases and pests in Africa: fungal and bacterial diseases. Acta Hort (213):151-159

Wakahiu MW, Gildemacher PR, Kinyua ZM, Kabira JN, Kimenju AW, Mutitu EW (2007) Occurrence of potato bacterial wilt caused by Ralstonia solanacearum in Kenya and opportunities for intervention. In: Khalf-Allah, AA et al (eds). Proceedings of the 7th Triennial Congress of the African Potato Association, October 22-26, 2007, Alexandria, Egypt, pp 267-271

Young N (1990) Seed potato systems in developed countries: Canada, The Netherlands and Great Britain. International Potato Center, Lima.

5 Improving seed health and seed performance by positive selection in three Kenyan potato varieties

Schulte-Geldermann, E.[a,‡], Gildemacher [a,b,c,‡], P.R., Struik, P.C.[c]

[a] International Potato Center (CIP), Nairobi, Kenya
[b] Royal Tropical Institute (KIT), Amsterdam, The Netherlands
[c] Centre for Crop Systems Analysis, Department of Plant Sciences, Wageningen University and Research Centre (WUR), Wageningen, The Netherlands
[‡] These authors contributed equally to the paper

Paper accepted for publication by the American Journal of Potato Research (13-04-2012)

Abstract

Selecting seed potatoes from healthy-looking mother plants (positive selection) was compared with common Kenyan farmer practice of selection from the harvested bulk of potatoes (farmer selection) in 23 farmer-managed trials. Positive selection assured lower incidences of PLRV (39%), PVY (35%) and PVX (35%). Positive selection out-yielded farmer selection irrespective of the agro-ecology, crop management, soil fertility, variety and quality of the starter seed, with an overall average of 30%. Regression analysis showed a relation between lower virus incidence and higher yield for the varieties. The paper discusses the consequences for seed system management in African countries. Furthermore possible additional effects of positive selection are discussed and further research is suggested. The paper concludes that positive selection can benefit all smallholder potato producers who at some stage select seed potatoes from their own fields, and should thus be incorporated routinely in agricultural extension efforts.

5.1 Introduction

The most important yield determining factor in potato cultivation is the quality of the seed tubers used (Lutaladio et al., 2009; Struik and Wiersema, 1999). The difficult availability of affordable high-quality seed potatoes is the major constraint for potato production systems worldwide (Lutaladio et al., 2009; Struik and Wiersema, 1999). Similarly, in East Africa the availability of affordable quality seed potatoes is a major obstacle for improving the profitability of potato production (Gildemacher et al., 2009b; Hirpa et al., 2010). The main factors reducing seed quality are biotic, including seed-borne viruses, bacteria and fungi, with a major component being the viruses (Salazar, 1996).

Positive seed potato selection is a simple technology to manage seed potato quality. It consists of selecting healthy-looking, vigorous mother plants to obtain seed tubers for the next seasons' crop. Positive selection has been proven to be a promising complementary technology for smallholder farmers in Kenya, in addition to seed production and marketing by specialized seed growers (Gildemacher et al., 2011). Specifically the fact that it has the potential to increase smallholder yields without monetary investment (Gildemacher et al., 2012b), and that it fits well into the prevailing seed sourcing strategy of smallholders, which is largely based on self- and neighbor supply (Gildemacher et al., 2009a) makes it a valuable complementary technology.

In earlier work (Gildemacher et al., 2011), a reduction in virus infection has been suggested as the main cause of yield increases by positive selection. This statement was supported by significant reduction in farmer-scored visual virus observations in farmer-managed demonstration trials, with no replications. The promising results have led to the incorporation of the technology as a component in seed system improvement interventions in Sub Sahara Africa (Gildemacher et al., 2012b). Considering this promotion of the technology through agricultural extension, it was deemed appropriate to further investigate the effect of positive selection on virus incidence in potato plant populations.

In this paper data is presented from replicated researcher managed and farmer managed trials comparing positive selection with seed recycling, which is the common farmer practice in Kenya (Gildemacher et al., 2009a). Virus incidence levels were scored through DAS-Elisa. This paper investigates the effect of positive selection on yield and virus incidence, and analyses the relation between virus incidence and yield. The paper discusses the proven and possible additional causes of the yield effect of positive selection, and provides suggestions for the direction of future research to optimize the use of the simple technology of positive selection in seed potato system improvement in Sub Sahara Africa.

5.2 Materials and methods

5.2.1 Source fields and starting material

Seed tuber lots were purposely planted for the experiments described in this paper. Seed potatoes were sourced from four different types of sources: 1) basic seed from KARI-Tigoni-3 field generations; 2) certified seed from ADC Molo-6 field generations; 3) potatoes sold as seed in rural markets – unknown number of generations; 4) farmer saved seed – unknown number of generations. Three different varieties were sourced: Dutch Robijn, Tigoni (CIP 381381.13) and Asante (CIP 381381.20). This resulted in a total of 23 different seed fields, planted without replication in a total of 15 locations.

Each of the 23 source fields was divided into two. From half the field, tubers were sourced by practicing positive selection, while on the other half of the field common farmer practice was applied. Positive selection entailed pegging healthy looking mother plants before full flowering by farmer groups which had received a basic group-based training (Gildemacher et al., 2007a). Just before full flowering virus infection symptoms are well visible. Two to four weeks after pegging the health status of the pegged plants was checked and pegs of plant showing newly developed systems removed. At harvest, pegged plants were harvested individually and seed sized tubers were collected to serve as seed tubers, provided all tubers of the individual plant looked healthy. Farmer selection consisted of the selection of seed sized tubers from the bulk of harvest potatoes from the other half of the source field, following common farmer practice.

5.2.2 Replicated farmer managed on-farm trials

The 23 sets of positive selection and farmer selection seeds were used to plant replicated farmer managed field trials at the same 15 different locations in the main potato growing areas of Kenya to compare the yields from the different types of seed tubers. Trials were planted either during the short rainy seasons (October '09 – January '10) or the long rainy season (April '10 – August '10), in a randomized complete block design (RCBD) layout with 3 or 4 replications of 40 plants each, at 30 cm x 75 cm distance.

Fertilization was based on 90 kg N/ha supplied in the form of NPK 10:26:10 at planting in the planting hole. Late blight was controlled through a spraying regime with Ridomil and

Mancozeb, adapted in response to actual disease occurrence. Further management was done according to farmer practice. At harvest all marketable tubers (> 30 mm) were collected and weighed.

For various reasons, e.g. porcupine damage, improper late-blight management or theft, not all 23 farmer managed experiments were successful. In total 21 data sets were obtained to reliably assess yield.

The pair-wise yield data was analyzed by testing for a significant difference between positive selection and farmer selection using a 1-tailed t-test.

5.2.3 Virus infection level testing

From the seed selected through positive selection and farmer selection random samples of 40 tubers were taken to assess virus infection, from 20 of the 23 trials. Plants were grown from eyes cut from the individual tubers and planted in aphid-free greenhouse chambers. After 4-6 weeks leaf sap was obtained from these plants (Casper and Meyer, 1981; Torrance, 1992). Leaf sap samples were tested individually for infection with PVY, PLRV and PVX through enzyme-linked immunosorbent assay (DAS-ELISA: CIP, Lima, Peru).

The pair-wise virus infection data (expressed in % for the individual viruses tested) and yield data were analyzed by testing for a significant difference between positive selection and farmer selection using 1-tailed t-tests. Virus infection levels of PLRV, PVY and PVX were plotted against yield and trend lines were fitted using the SPSS curve estimation procedure (IBM SPSS statistics 20). In addition a combined virus infection indicator was calculated by taking the simple absolute sum of the number of infected sample tubers in the total sample of 40 tubers per treatment. This index can be larger than the total number of tubers in the sample and should be considered as a measure for the virus load. For this index the same plotting procedure was used.

5.2.4 Replicated on-station factorial fertilizer x seed source trials

An additional replicated factorial trial, with the varieties Asante, Tigoni and "Purple Tigoni", a landrace, was planted under researcher control to test for fertilizer seed selection interaction during the long rainy season of 2010. The trial was planted in a split-plot layout with fertilizer treatments as main plots and selection method x variety as subplots. The seed was sourced from four ware potato farmer groups involved in the on-farm trials described above. Two fertilizer levels were implemented using NPK 10:26:10 at 45 and 90 kg/ha N equivalent. Each plot counted 40 tubers at 30 cm within-row and 75 cm between-row distance. At harvest all marketable tubers (> 30 mm) were collected, weighed and counted. Data were analyzed with ANOVA and protected LSD values were assessed.

5.3 Results

The results from a total of 18 farmer managed randomized complete block trials are summarized in Table 5.1. When analyzing the combined paired observations through a t-test, positive selection clearly out-yielded the farmer selection treatment, irrespective of variety or quality of the starter material.

For all three varieties, positive selection resulted in substantially higher yields, the difference ranging from an average of 23% for Tigoni to 35% for Dutch Robijn. The average absolute yield increase for high-quality starting material was 7.3 t/ha, whereas farmer-quality starter material gave an average yield increase of 3.0 t/ha.

Table 5.1: Yield from positive seed potato selection compared with common farmer seed potato selection presented separately for the varieties Asante, Tigoni and Dutch Robijn, and for two quality levels of the source field

Name	Yield[a]				Yield increase				
	PS (t/ha)	Std. Dev.	FS (t/ha)	Std. Dev.	(t/ha)	%	t-value	df	p (2-tailed)
Asante	13.9	1.97	11.0	2.20	2.8	25	7.49	21	0.000
Tigoni	18.6	7.37	15.1	5.99	3.5	23	3.75	19	0.001
Dutch Robijn	20.9	10.13	15.5	6.93	5.4	35	6.31	25	0.000
Farmer quality[b]	14.1	2.31	11.1	2.11	3.0	27	4.84	51	0.000
High quality[c]	30.5	6.99	23.1	4.19	7.3	32	13.36	15	0.000
All	17.9	8.00	13.9	5.80	4.0	30	14.38	67	0.000

[a] PS= positive selection; FS= farmer selection
[b] Seed potatoes for these trials selected from fields planted with seed potatoes from local market or farmer fields
[c] Seed potatoes for these trials selected from fields planted with basic seed or certified seed

Table 5.2 shows the summarized results of the researcher-controlled replicated trials implemented with seed potatoes sourced through positive selection and through farmer selection. Two different fertilizer regimes were implemented to assess the interaction with the effect of positive selection. There was no interaction between fertilizer level and the effect of positive selection. A higher fertilizer level did increase significantly both yield and the number of tubers harvested. Positive selection increased the average yield to 13.4 t/ha as compared with 10.6 t/ha for using farmer selection seed potatoes. This constitutes a 26% increase in yield. The number of tubers harvested increased by 23%, from 21.9 to 26.8 m^{-2}.

Table 5.3 shows the average effect of positive selection compared with farmer selection on PLRV, PVY and PVX infection rates as measured through DAS Elisa testing. The results show a significant reduction in PLRV, PVY and PVX infection rates. The average measured PVY infection was 25.0% after positive selection, compared with 38.4% for farmer selection. The infection rate with PLRV went down from 31.1% for farmer selection to 19% for positive selection. At the same time yields increased by an average 25.6% as a result of positive selection.

Table 5.2: Effect of seed selection method under two different fertilizer regimes (45 or 90 kg N per ha based on NPK 10-26-10) on yield (t/ha) and number of tubers (#/m2), Kabete, Nairobi, Kenya, 2010

Fertilizer regime and seed selection method	Yield[a] (t/ha)	No. tubers[b] (#/m^2)
45 kg N/ha		
Positive selection	11.7	23.8
Farmer selection	8.9	18.7
Total	**10.3**	**21.3**
90 kg N/ha		
Positive selection	15.0	29.8
Farmer selection	12.4	25.0
Total	**13.7**	**27.4**

[a] No significant interaction between fertilizer level and selection method (F pr. =0.698). All yield contrasts significantly different, LSD 5%= 0.49
[b] No significant interaction between fertilizer level and selection method (F pr. =0.759). All yield contrasts significantly different, LSD 5%= 1.30

Table 5.3: Effect of positive selection compared with farmer selection on the level of infection with potato viruses PLRV, PVX and PVY and progeny yield (t/ha) of selected seed tubers.

	Farmer selection (FS)		Positive selection (PS)		Difference PS-FS			
	Infection rate (%)	Std. dev.	Infection rate (%)	Std. dev.	Infection rate (%)	Relative change (%)	t value (df = 20)	P (1-tailed)
PLRV	31.1	14.15	19.0	8.17	−12.1	39.0	7.00	0.0000
PVX	7.5	6.54	4.9	4.55	−2.6	35.0	2.93	0.0043
PVY	38.4	18.40	25.0	12.57	−13.4	34.9	8.06	0.0000
	Yield (t/ha)	Std. dev.	Yield (t/ha)	Std. dev.	Yield (t/ha)	Relative change (%)	t value (df = 20)	P (1-tailed)
Tuber fresh yield	12.90	4.72	16.21	5.81	3.31	25.6	7.72	0.0000

The yield differences observed in the individual trials were plotted against the virus infection rate (Figure 5.1). For Tigoni and to a somewhat lesser extent Dutch Robijn the expected negative correlation between yield and virus levels was observed. The variety Asante did not respond significantly to a reduction in infection with PVY and PVX, and responded less than the other two varieties (but significantly so) to a reduction in PLRV. When plotting tuber yield against the sum of the number of infected tubers in each sample (assessed through growing plants from eye-cuttings and testing virus infection in a leaf sample) for the three viruses, this could account for 70% and 46% of the variation for Tigoni and Dutch Robijn, respectively, while for Asante it could only account for 17% of the variation.

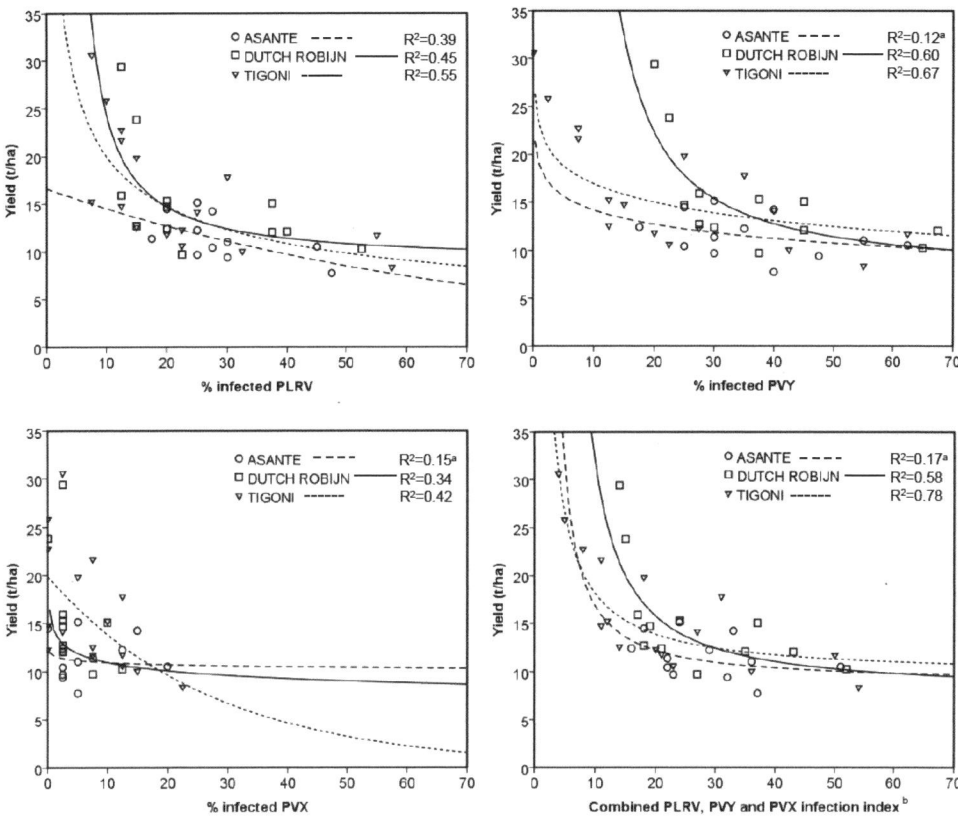

Figure 5.1: Fresh tuber yield ("Yield") plotted against the percentage infection with PVY, PVX or PLRV and against the sum of the infection count of all three viruses for the potato varieties Asante, Dutch Robijn and Tigoni.

[a] ns = non significant regression
[b] index calculated as the simple absolute sum of the number of infected tubers for each virus in the total sample of 40 tubers. This index can be larger than the total number of tubers in the sample and should be considered as a measure for the virus load.

5.4 Discussion

5.4.1 Positive selection results in yield increase compared with farmer selection

The results from the trials confirm the earlier findings from farmer managed demonstration trials (Gildemacher et al., 2011; Gildemacher et al., 2012b) that positive selection is a technology that can provide smallholder producers with a significant yield advantage compared with the common practice of indiscriminate recycling of seed potatoes from their ware potato harvest. For three different popular Kenyan potato varieties over a wide range of agro-ecologies and (farmer) management practices (Table 5.1), for two levels of fertilizer application (Table 5.2) and for different qualities of primary material (Table 5.1), the simple practice of selecting the best looking plants as a seed source for next season's crop gave a highly significant yield advantage under all tested circumstances over just selecting from the bulk of the harvest, as farmers tend to do (Gildemacher et al., 2009a).

The three different varieties appeared to differ in their reaction to positive selection, with Dutch Robijn responding most strongly with an average yield advantage of positive selection over common farmer practice of 5.4 t/ha, or 35%, whereas for Tigoni and Asante varieties a yield increase of 23 and 25% was recorded on average (Table 5.1).

The farmer managed trials showed that positive selection is an effective technology for seed potato populations of different generations. Positive selection resulted in substantial yield increases when used to source seed potatoes from farmer ware potato plots, but it also increased yields when applied on fields planted with relatively high-quality seed potatoes from certified seed potatoes (6 generations of multiplication) and basic seed potatoes (3 generations of multiplication) (Table 5.1).

It is in itself not surprising that the technology works, as it has been in use in clonal selection in the first stages of conventional seed potato multiplication (Salazar, 1996; Struik and Wiersema, 1999). The Canadian seed potato system even has a specific directive describing the accepted procedure to source starter material for pre-elite seed production from ordinary potato fields through positive selection (Canadian_Food_Inspection_Agency, 2010). However, what is more important is that also ware potato producers are able to learn and implement the technology successfully (Bryan, 1983) and substantially increase their productivity compared with their common practice of selecting seed potatoes from the bulk of the harvested tubers.

5.4.2 Positive selection reduces virus infection compared with farmer practice

Explaining the convincingly demonstrated yield increases is however more complicated. The combined results of the farmer managed and researcher managed trials demonstrated that positive selection had a strong effect on the measured virus infection rate of seed tubers selected through positive selection compared to seed tubers selected from the bulk of the harvest (Table 5.3; Figure 2.1). PVY and PLRV infection rates measured in the farmer selection seed were above 30%, while in the positive selection treatment the infection rate was reduced by 35 and 39%, respectively (Table5.3). The measured rates of infection for PVX in farmer selection was lower at 7.5%, but also PVX levels were reduced after positive selection as compared with farmer selection (Table 5.3). Even under the high disease pressure prevailing in farmer managed fields with potatoes that were recycled for several generations, substantial reductions were measured in the infection rates of the selected tubers. This shows that positive selection, practiced by smallholder ware potato producers is a powerful technology to keep virus infection levels in check.

5.4.3 The relationship between yield increase and virus reduction

Both yield increases and a reduction in virus infection could be demonstrated beyond any doubt in a very wide variety of circumstances. In addition the data show a significant correlation between yield and virus infections (Figure 5.1), suggesting that an important mechanism behind the effect of positive selection is the reduction in virus infection in the plant population.

For Tigoni and Dutch Robijn, a similar response can be observed (Figure 5.1). Lower yields were obtained when infection rates were higher. The figure demonstrates graphically that in the trials in which higher yields were obtained, an increase in virus incidence was of more consequence compared to the trials in which low yields were obtained. This is in line with the observation in Table 5.1 that positive selection gives high yield increases when applied on fairly high-quality seed.

The correlation between PLRV, PVY and PVX infection and yield of Tigoni and Dutch Robijn is significant with respectively 12 and 16 data points (Figure 5.1) under highly heterogeneous circumstances. Seed potatoes were derived from different sources with widely different history in terms of number of generations of multiplication and crop management. In addition the trials were planted under different circumstances across Kenya, with as a consequence different virus pressure. Finally the actual positive selection was done by different groups of producers. Still, a convincing correlation could be demonstrated between virus infection rates and yield.

The shape of the relationship between virus infection and yield is different from what could be expected based on work by Reestman, who predicts an S-shaped curve, with limited effects of virus infection at lower incidences, and a diminishing effect at very high incidences (Reestman, 1946; Reestman, 1970), based on the compensation effect. Van der Zaag (1987) calculated that further increasing virus incidence would have stronger yield consequences in an already highly infected crop compared to a lightly infected crop. It is mentioned by Reestman (1970) that the yield reducing effects mentioned across the literature of that time are diverse and confusing, and he suggested that this depended on the rate of compensation by neighboring plants. Low fertility, drought and few stems per plant were mentioned as factors impeding compensation. All these three factors are very common in smallholder potato production in Kenya. Potato virus infection may well do more damage under these sub-optimal growth conditions than under more favorable conditions. Furthermore infection rates under which virus effects were observed to be largely compensated by neighboring healthy plants were low (<10%) (Reestman, 1970). Authors agree that damage can become more severe as the result of multiple infections with different viruses, a situation common in Kenya (Gildemacher et al., 2009a; Radcliffe and Ragsdale, 2002).

5.4.4 Possible additional mechanisms contributing to the effect of positive selection

The variety Asante only showed a significant yield response to a reduction in PLRV infection, but less pronounced than the varieties Tigoni and Dutch Robijn. In addition Asante showed no response to a reduction in PVY or PVX. This suggests that Asante harbors partial resistance or tolerance for these viruses. Differences in response to virus infection between varieties are common (Bawden et al., 1948; de Bokx, 1972; MacKinnon and Munro, 1959; Radcliffe and Ragsdale, 2002). In the case of extreme resistance however, the virus would not be detected through DAS-ELISA.

The yield response of Asante to positive selection was similar as for the other two varieties, while it does respond much less strongly to a reduction in PVY, PLRV and PVX incidence. This suggests that in addition to a lower incidence of PLRV, PVY and PVX, there might have been other factors that played a role in the measured effects on yield resulting from positive selection.

The most likely explanation is that other viruses which were not tested have also been reduced in incidence as a result of positive selection. PVM and PVS were found to be abundant in farmer based seed systems in Kenya (Muthomi et al., 2009), while PVA was found to be common in seed tubers sold at Kenyan rural markets (Gildemacher et al., 2007b). PVA is often not considered to cause serious damage, but can cause severe symptoms in combination with PVY or PVX (Nganga and Shideler, 1982), a common combination of infections in farmer fields in Kenya (Gildemacher et al., 2009a). Any endemic virus disease that would cause visible symptoms will be affected by positive selection, and would have contributed to the yield increase realized by positive selection compared with common farmer practice.

Synergistic effects of infection with multiple viruses have been described by several authors for different virus combinations of sweet potato (Karyeija et al., 2000; Untiveros et al., 2007) soybean (Malapi-Nelson et al., 2009), tomato (Balogun et al., 2005), wheat (Tatineni et al.), and numerous other crops. In case of synergistic effects between viruses on potato yield, this may further increase the influence of a reduction in virus incidence of those viruses not detected here. Synergistic effects between PLRV and both PVY and PVX were demonstrated for susceptible potato varieties (Brandolini et al., 1992).

Similarly to virus infection levels, the levels of other seed borne diseases having an effect on general plant appearance will be affected by positive selection. Turkensteen (1987) identified *Erwinia* spp. bacteria (currently *Pectobacterium* spp.) and *Fusarium* spp. fungi as 'important' seed-borne pathogens in Central Africa. In addition bacterial wilt (*Ralstonia solanacearum*) is endemic in Kenya (Wakahiu et al., 2007). For the latter however the symptoms can hardly be mistaken and all was done to avoid bacterial wilt infection of the trials.

In addition to having an effect on the incidence of viruses on the potato plant population, it cannot be ruled out that positive selection results in a lower virus load of individual seed potatoes, the virus titer. Little is known about the effect of virus titer in seed tubers on the final yield. Van der Zaag (1987) reported that tubers infected late in the previous season had less severe symptoms and yield reduction than those infected earlier and mentioned that diseased tubers that had been recycled for a number of generations did worse than those having a shorter history of infection. However, no data to support this were presented. Barker and Woodford (1987) reported unusually mild PLRV symptoms in the progeny of late-infected mother plants. Interesting enough however, they could not show a difference in virus titer in the leaves about 7 weeks after planting the progeny tubers. Satoh et al. (2011) do provide some evidence that the gene response and resulting symptom expressions of rice plants to infection with Rice Dwarf Virus (RDV) are related to the concentration of the virus in the plant. Further research to improve the understanding of the effect of positive selection on average potato crop virus titer, and the effect of virus titer on crop growth would be of interest.

5.4.5 Effect of soil fertility on effectiveness of positive selection

Not surprisingly potato yields increased with a higher fertilizer application. It could theoretically be expected that poorly nourished plants suffer more from the same level of virus infection, as has also been reported (Bokx and Want, 1987). Especially abundant nitrogen fertilizer can mask the visual mosaic symptoms of virus infection (Salazar, 1996). On this basis it could be hypothesized that a poorly nourished crop would benefit more from positive selection than a well fertilized crop. However, under the fertilizer regimes tested no interaction between the effect of positive selection and soil fertility management could be observed.

5.4.6 Remaining research questions

The result from this research has conclusively shown that positive selection is a suitable technology for seed potato quality maintenance by smallholder potato producers. The viruses PVX, PVY and PLRV have shown to play a role, but it appears there may be additional factors, most likely other viruses and possibly other soil borne diseases, contributing to the effect of the technology. Targeted controlled research to investigate such additional factors would increase the understanding of the mechanisms behind positive selection.

What makes even more curious is the full potential of positive selection over several generations. Current seed potato system management decisions are based on the assumption that degeneration as a result of tuber-borne diseases is an inevitable fact, and that regular seed renewal from a reliable disease-free source is the only manner to maintain an acceptable yield potential. This research has shown that positive selection assists in managing virus infection levels. It would be of interest to witness potato yields over several generations of applying positive selection to a degenerated potato crop. This would allow one to challenge the common believe that degeneration is inevitable and irreversible in a potato population. It could be hypothesized that opposed to degeneration of a potato plant population over generations, also regeneration needs to be considered an option, provided ware potato farmers manage their selection process well.

Answering this question is of great essence for seed potato systems in countries where production levels are, unlike in some developed countries, not close to the theoretically optimal production. A better understanding of the rate of degeneration in relation to disease pressure and farmer management will allow for better informed investments in seed potato program building and the seed renewal strategy by individual potato producers. Combined economic and seed degeneration research could contribute to this improved decision making.

Virus resistance and virus tolerance are elements requiring attention in further research. Genetic variation in virus resistance and tolerance has been identified and is used in breeding programs (Arif et al., 2011; Brandolini et al., 1992; Munro, 1961). Combining virus resistance in popular potato varieties with better seed quality management by ware potato producers through positive selection may reduce the importance of commercial seed potato multiplication, which has proven to be difficult to establish in developing countries.

Before such far reaching conclusions can be made, however, further follow-up research to address the point above is needed. The authors would, in this regard, like to suggest a factorial trial with variety (1), starter seed infection (2), and seed quality management (3) as factors, to be continued over a minimum of four generations:
1. Variety
    a. Fully susceptible variety
    b. Best known resistant and tolerant variety
2. Starter seed infection
    a. Tested and quantified highly virus infected seed potatoes
    b. Tested and quantified very low virus infected seed potatoes
3. Seed quality management
    a. Seed recycling with positive selection
    b. Blind seed recycling (farmer selection)
    c. Flush out by new seed multiplied from original starter seed each season

Virus infection levels would have to be monitored intensively to assure maximum understanding of the virus epidemiology, and especially the dynamics of virus epidemiology over generations, both in terms of the fractions of infected plants and tubers, but also in terms of titer build-up in plants and tubers.

Such research is time and resource consuming and risk prone. Frequent observations need to be made and samples taken and investigated over a number of generations, which would call for trials, situated on research stations with reliable irrigation systems. However, to stay close to common practice, on-farm trials may be better suited, which would reduce researcher control and increase the risk of failure somewhere over the four generations.

Finally, as (Döring, 2011) indicates, virus epidemiology is highly complex as a result of the numerous interactions between viruses, vectors, the plant and the environment. He observed that to detect meaningful patterns large amounts of data are needed, after which the question remains to what extent findings can be generalized. Still, a better understanding of how positive selection impacts on yield of potatoes would in our case be supportive to efforts to make the technology standard practice of potato producers who have the habit to recycle seed from their last crop.

5.4.7 Consequences of the research findings

The research findings demonstrate conclusively that positive selection is a technology that works under very diverse circumstances. If potato producers decide to source seed potatoes from their previous crop, rather than renewing their seed from a reliable source of high-quality seed, positive selection is highly recommended.

Considering the highly conclusive results with regard to the effect of positive selection, and the fact that the technology requires only sticks or another type of marker and labor as input, it is very suitable for seed quality maintenance by smallholder potato farmers, who form the majority of potato farmers in Sub-Sahara Africa. Gildemacher et al. (2011) demonstrated that an average Kenyan potato producer who invests 4 mandays per hectare in positive selection can realize 284 Euro additional profit, after deducting 6 Euro opportunity cost for his labor investment. In addition positive seed potato selection can fairly easily be learned (Gildemacher et al., 2012a). Based on these findings it can be recommended to include positive selection in the training curricula and programs of smallholder potato farmers in countries where sourcing seed potatoes from their own ware potato crop is common practice.

It has to be emphasized that positive selection is not very suitable for commercial seed potato multiplication. For seed quality maintenance it would suffice to mark roughly 10%-15% of the potato plants in a field as mother plants to source seeds for the next season for a same sized plot. A seed multiplier requires to bulk seed over seasons from a limited amount of starter seed, and thus requires harvesting the vast majority of his plants as seed. For seed multiplication removing visibly infected plants, negative selection, remains the only option.

References

Arif, M., U. Azhar, M. Arshad, Y. Zafar, S. Mansoor, and S. Asad. 2011. Engineering broad-spectrum resistance against RNA viruses in potato. Transgenic Research:1-9.

Balogun, O.S., T. Teraoka, and Y. Kunimi. 2005. Influence of the host cultivar on disease and viral accumulation dynamics in tomato under mixed infection with Potato virus X and Tomato mosaic virus. Phytopathologia Mediterranea 44:29-37.

Barker, H., and J. Woodford. 1987. Unusually mild symptoms of potato leafroll virus in the progeny of late-infected mother plants. Potato Research 30:345-348.

Bawden, F.C., B. Kassanis, and F.M. Roberts. 1948. Studies on the importance and control of Potato Virus X. Annals of Applied Biology 35:250-265.

Brandolini, A., P.D.S. Caligari, and H.A. Mendoza. 1992. Combining resistance to potato leafroll virus (PLRV) with immunity to potato viruses X and Y (PVX and PVY). Euphytica 61:37-42.

Bryan, J. 1983. On-farm seed improvement by the potato seed plot technique. International Potato Center, Lima, pp. 13.

Canadian_Food_Inspection_Agency. 2010. Requirements for the production of Pre-Elite seed potatoes from sources other than Nuclear Stock. In C. F. I. Agency, (ed.) Directive D-97-11, 2nd revised ed. Canadian Food Inspection Agency -, Ottawa, Ontario.

Casper, R., and S. Meyer. 1981. Die Anwendung des ELISA-Verfahrens zum Nachweis pflanzenpathogener Viren. Nachrichtenblatt des Deutschen Pflanzenschutzdienstes: 33:49-54.

De Bokx, J.A. 1972. Spread of potato virus S. Potato Research 15:67-70.

De Bokx, J., and J. van de Want. 1987. Viruses of potatoes and seed-potato production. 2 ed. Pudoc, Wageningen.

Döring, T.F. 2011. Potential and limitations of plant virus epidemiology: Lessons from the Potato Virus Y pathosystem. Potato Research 54:341-354.

Gildemacher, P., P. Demo, P. Kinyae, M. Wakahiu, M. Nyongesa, and T. Zschocke. 2007a. Select the best: positive selection to improve farm saved seed potatoes; Trainers manual. International Potato Center, Nairobi.

Gildemacher, P.R., J. Mwangi, P. Demo, and I. Barker. 2007b. Prevalence of potato viruses in Kenya and consequences for seed potato system research and development, pp. 84-92, In A. Khalf-Allah (ed.), 7th Triennial African Potato Association conference. African Potato Association, Alexandria, Egypt.

Gildemacher, P., P. Demo, I. Barker, W. Kaguongo, G. Woldegiorgis, W. Wagoire, M. Wakahiu, C. Leeuwis, and P.C. Struik. 2009a. A description of seed potato systems in Kenya, Uganda and Ethiopia. American Journal of Potato Research 86:373-382.

Gildemacher, P., W. Kaguongo, O. Ortiz, A. Tesfaye, G. Woldegiorgis, W. Wagoire, R. Kakuhenzire, P. Kinyae, M. Nyongesa, P.C. Struik, and C. Leeuwis. 2009b. Improving potato production in Kenya, Uganda and Ethiopia: a system diagnosis. Potato Research 52:173-205.

Gildemacher, P., E. Schulte-Geldermann, D. Borus, P. Demo, P. Kinyae, P. Mundia, and P.C. Struik. 2011. Seed potato quality improvement through positive selection by smallholder farmers in Kenya. Potato Research 54:253-266.

Gildemacher, P., C. Leeuwis, P. Demo, P. Kinyae, P. Mundia, M. Nyongesa, and P. Struik. 2012. Dissecting a successful research-led innovation process: The case of positive seed potato selection in Kenya. International Journal of Technology Management and Sustainable Development. SUBMITTED.

Hirpa, A., M. Meuwissen, A. Tesfaye, W.J.M. Lommen, A. Oude Lansink, A. Tsegaye, and P.C. Struik. 2010. Analysis of Seed Potato Systems in Ethiopia. American journal of potato research 87:537-552.

Karyeija, R.F., J.F. Kreuze, R.W. Gibson, and J.P.T. Valkonen. 2000. Synergistic interactions of a Potyvirus and a phloem-limited Crinivirus in sweet potato plants. Virology 269:26-36.

Lutaladio, N., O. Ortiz, A. Haverkort, and D. Caldiz. 2009. Sustainable Potato Production; guidelines for Developing Countries FAO, Rome.

MacKinnon, J., and J. Munro. 1959. Comparative rates of movement of potato virus X into tubers and eyes of three potato varieties. American Journal of Potato Research 36:410-413.

Malapi-Nelson, M., R.H. Wen, B.H. Ownley, and M.R. Hajimorad. 2009. Co-infection of soybean with Soybean mosaic virus and Alfalfa mosaic virus results in disease synergism and alteration in accumulation level of both viruses. Plant Disease 93:1259-1264.

Munro, J. 1961. The importance of potato virus X. American Journal of Potato Research 38:440-447.

Muthomi, J., J. Nyaga, F. Olubayu, J. Nderitu, J. Kabira, S. Kiretai, J. Auro, and M. Wakahiu. 2009. Incidence of aphid-transmitted viruses in farmer-based seed potato production in Kenya. Asian Journal of Plant Sciences 8:166-171.

Nganga, S., and F. Shideler. 1982. Potato Seed Production for Tropical Africa International Potato Center, Nairobi.

Radcliffe, E., and D. Ragsdale. 2002. Aphid-transmitted potato viruses: The importance of understanding vector biology. American Journal of Potato Research 79:353-386.

Reestman, A.J. 1946. De beteekenis van de virusziekten van de aardappel naar aanleiding van proeven met gekeurd en ongekeurd pootgoed. European Journal of Plant Pathology 52:97-118.

Reestman, A.J. 1970. Importance of the degree of virus infection for the production of ware potatoes. Potato Research 13:248-268.

Salazar, L. 1996. Potato viruses and their control International Potato Center (CIP), Lima.

Satoh, K., T. Shimizu, H. Kondoh, A. Hiraguri, T. Sasaya, I.R. Choi, T. Omura, and S. Kikuchi. 2011. Relationship between symptoms and gene expression induced by the infection of three strains of Rice dwarf virus. PLoS ONE 6.

Struik, P.C., and S.G. Wiersema. 1999. Seed potato technology Wageningen University Press, Wageningen.

Tatineni, S., R.A. Graybosch, G.L. Hein, S.N. Wegulo, and R. French. Wheat cultivar-specific disease synergism and alteration of virus accumulation during co-infection with wheat streak mosaic virus and Triticum mosaic virus. Phytopathology 100:230-238.

Torrance, L. 1992. Developments in methodology of plant virus detection. Netherlands Journal of Plant Pathology 98:21-28.

Turkensteen, L.J. 1987. Survey of diseases and pests in Africa: fungal and bacterial diseases. Acta Horticulturae:151-159.

Untiveros, M., S. Fuentes, and L.F. Salazar. 2007. Synergistic interaction of Sweet potato chlorotic stunt virus (Crinivirus) with carla-, cucumo-, ipomo-, and potyviruses infecting sweet potato. Plant Disease 91:669-676.

Van der Zaag, D.E. 1987. Growing seed potatoes, p. 176-203, In J. de Bokx et al. (eds.), Viruses of potatoes and seed-potato production, 2nd edition. Pudoc, Wageningen.

Wakahiu, M.W., P.R. Gildemacher, Z.M. Kinyua, J.N. Kabira, A.W. Kimenju, and E.W. Mutitu. 2007. Occurrence of potato bacterial wilt caused by Ralstonia solanacearum in Kenya and opportunities for intervention., pp. 267-271. In A. Khalf-Allah (ed.), 7th Triennial African Potato Association Conference. African Potato Association, Alexandria, Egypt.

6 Dissecting a successful research-led innovation process: the case of positive selection in seed potato production in Kenya

Gildemacher, P.R.[a,b], Leeuwis, C.[c], Demo, P.[a], E.[a], Kinyae, P.[c], Mundia, P.[d], Nyongesa, M.[d], Struik, P.C.[c]

[a] International Potato Center (CIP), PO Box 25171, Nairobi, Kenya
[b] Royal Tropical Institute (KIT), PO Box 95001, 1090 HA, Amsterdam, The Netherlands
[c] Wageningen University and Research Centre (WUR), PO Box 9101, 6700 HB, Wageningen, The Netherlands
[d] Kenya Agricultural Research Institute (KARI), Tigoni, PO Box 338, Limuru, Kenya
[e] Jomo Kenyatta University of Agriculture and Technology (JKUAT), PO Box 62000, Nairobi Kenya

Submitted to: International Journal of Technology Management and Sustainable Development (18-04-2012)

Abstract

The role and position of agricultural research in innovation for development are subject to debate. By identifying the success factors of a research-led programme on positive seed potato selection, the role of research in agricultural innovation is analysed. With minimum resources, the positive seed selection programme developed an approach to improve the quality of seed potatoes by ware potato growers, complementary to specialised seed production systems and now widely promoted in sub-Sahara Africa. The case points out that innovation can emerge from old technology within existing institutional environments. Placing innovation central, rather than research outcome, widens the role of research. In our specific case, research assumed responsibility, in partnership with extension, for developing and piloting effective training. Researchers effectively contributed to innovation because they were given and took the liberty of pursuing a 'bright idea'. It proves worthwhile to search for opportunities to innovate without complex institutional change. The positive selection case shows that these opportunities can be surprisingly simple as long as researchers have room to manoeuvre and opportunity to immerse in practical collaborative partnerships with practitioners.

6.1 Introduction

Innovation is an important driver for economic development, productivity growth and increased welfare (Edquist and Johnson, 1997; Lundvall, 1992). The understanding of the role of research in the process of innovation has changed over time. The linear model of thinking with research as the main driver and unique initiator of innovation has been dismissed (Arnold and Bell, 2001), as it does not do justice to the interplay between science, technology, economic actors and society at large and the multiple sources of innovation. Many authors advocated for a shift towards innovation system thinking (Arnold and Bell, 2001; Hall et al., 2001; Spielman et al., 2009; Woolthuis et al., 2005), which places the interaction between diverse actors, including the private sector, central in the process of innovation (Biggs, 2007; Engel, 1995; Hall, 2006; Hall, 2007). Furthermore its central focus is putting knowledge into use, rather than research output.

Technical innovation can often only be realized in combination with changes in the social-organizational sphere (Leeuwis, 2004; Worldbank, 2007a), such as new land-tenure arrangements (Adjei-Nsiah et al., 2004; Dormon et al., 2007), price incentives (Dormon et al., 2007), new market arrangements (KIT and CFC, 2011). Such institutional innovations are no longer considered external conditions influencing adoption of technology, but rather integral parts of an innovation. Therefore, innovation has been re-conceptualized as a successful combination of 'hardware' (new technical devices and practices), 'software' (new knowledge and modes of thinking) and 'orgware' (new social institutions and forms of organization) (Smits, 2000). This re-conceptualization reinforces the idea that effective innovation processes require many stakeholders and networks to be involved in design and application.

Agricultural research has an important role to play in innovation. The role, mandate and position of researchers and research organizations within such a broader concept of innovation remains a point of debate (Worldbank, 2007a), a debate that has evolved since the 1980s (Chambers, 1983). Somewhat confined but commonly accepted roles of research in innovation are: developing entry points and ideas for change based on existing and new knowledge and leading the process of structured testing and adaptation of new technical practices and approaches, thus allowing for objective decision making. However, researchers and research organizations could also facilitate alignment with the 'software' and 'orgware' dimensions of innovation, by articulating societal knowledge demands, engaging in participatory research, building multi-stakeholder networks, facilitating learning, and designing mechanisms to make research more accountable to society (Funtowicz and Ravetz, 1993; Leeuwis and Aarts, 2011). Such tasks fit under the banner of 'innovation intermediation' (Howells, 2006; Klerkx and Leeuwis, 2008a), 'innovation broking' (Klerkx et al., 2009) or 'innovation management' (Schut et al., 2011).

While numerous research organizations aim to play such broader roles, it is widely acknowledged that ingrained and institutionalized procedures and cultures in research establishments are not always conducive to putting such ideas in practice (Devaux et al., 2009; Hocdé et al., 2008; Kristjanson et al., 2009; Rajalahti et al., 2008; Worldbank, 2007a).

This paper aims to contribute to the understanding of the effective contribution of agricultural research to innovation processes by presenting a case in which researchers realized innovation in East Africa. It consists of an intervention to pilot positive seed potato selection as a means for Kenyan ware potato producers to improve the quality of their planting material. Positive selection can be considered as an old and technically inferior technology. It is so simple that one wonders why it is not used already. The practice of seed potato quality management by ware potato farmers is radically different from decades of research and development efforts focusing almost exclusively on seed quality improvement in developing countries through specialized multiplication systems.

Positive selection proved very effective in increasing smallholder potato yields without any cash investment. The training method developed resulted in rapid adoption. The training manual has been translated in seven languages and is currently used in at least 10 countries. To date over 43,000 producers have been trained in Kenya, Rwanda, Uganda, Rwanda and Malawi alone. Finally the initiative has resulted in the acceptance of the improvement of seed quality management by ware potato producers as an important additional approach to improving the seed potato quality of smallholders in Sub-Saharan Africa (Lutaladio et al., 2009).

This intriguing example of an innovation process triggers reflection on the role of research and research organizations. We first present the technology, the results achieved and the success factors of the intervention. Then we reflect on the extent to which newly advocated roles for research contributed to the successful innovation, and on how this was supported or hindered by dominant modes of thinking and doing within the research organization. Lessons will be drawn regarding the roles of research(ers) in agricultural innovation, and the conditions that favour an effective fulfillment of such roles.

6.2 Positive seed potato selection

6.2.1 Seed potato systems in Kenya

Potato (*Solanum tuberosum* L.) is an important staple and cash crop for smallholder farmers in the Kenyan highlands. Poor quality of seed tubers is a major yield-reducing factor in potato production in Eastern Africa (Gildemacher et al., 2009a; Gildemacher et al., 2009b). High virus incidences in seed tubers greatly contribute to prevailing poor yields (Muthomi et al., 2009). Improving seed potato quality is considered a pathway to improve smallholder potato yields and income (Eshetu et al., 2005; Getachew and Mela, 2000; Muthomi et al., 2009; Tindimubona et al., 2000).

Over generations seed tuber quality degenerates as tuber-borne diseases, mainly viruses, accumulate, causing the yield potential of the seed to diminish (Salazar, 1996; Struik and Wiersema, 1999). Yield loss can be avoided by regularly replenishing seed stocks by high-quality seed potatoes with little virus infection. The specialized production, distribution and quality control system required make high-quality seed potatoes inaccessible to almost all potato producers in Eastern and Southern Africa. Currently, the seed potato systems in the African highlands are dominated by neighbour and self-supply (Crissman et al., 1993; Gildemacher et al., 2009a), with little value difference between seed and ware tubers.

Seed system interventions to improve smallholder potato yields have invariably been focussed on specialised seed potato growers, supplying smallholder ware potato farmers. Notwithstanding pilot successes there is little evidence of drastic and sustainable improvement of the yields of poor potato producers in Sub-Saharan Africa as a result of these efforts (Gildemacher et al., 2009a). Considering the importance of farm-saved seed potatoes in Eastern Africa, the need was identified to improve seed potato quality management by ware potato producers as an additional strategy to improving the overall quality of seed potatoes used (Gildemacher et al., 2009a). This raised the following questions: 1) what technologies can smallholder ware potato growers apply to maintain or even improve the quality of their seed potato stocks?, and 2) how can these technologies be made available to many smallholder ware potato growers?

6.2.2 Positive seed potato selection

Positive seed potato selection was identified as a technology that could assist potato producers in Kenya. It is based on the following actions: 1. the best potato plants in a field are identified and marked before crop senescence starts and obscures disease symptoms; 2. the marked plants are harvested separately as seed source; 3. these seed tubers are stored separately. Originally, positive selection was used primarily in formal seed potato multiplication to select disease-free mother plants as the starting point of clonal selection (Bokx and Want, 1987; Salazar, 1996). It has been used in Central Africa to start up a seed multiplication system (Haverkort, 1986). Currently, seed tubers in formal seed systems are mostly multiplied from tested, disease–free, tissue-culture material or from other disease-free nuclear stock. On-farm positive selection to maintain seed potato

6 Dissecting a successful research-led innovation process

quality is also mentioned in literature (Struik and Wiersema, 1999), and in instruction materials (Bryan, 1983), but is not commonly used by ware potato producers (Gildemacher et al., 2009a; Hirpa et al., 2010), nor was it promoted in any potato development programme known to the authors.

Box 6.1 Positive selection trial set-up (based on Gildemacher et al., 2011).
Reproduced with kind permission from Springer Scientific Publishers and the EAPR.

1. Let the farmer group select an average potato field.
2. Divide it into two and let farmers peg healthy looking plants just before flowering in half of the field; reconfirm the health status of pegged plants two weeks later.
3. Harvest seed potatoes after judging the tubers of each pegged plant in the positive selection plot; select seed from the farmer practice plot using common farmer practice.
4. Store seed potatoes from both sources under the same conditions.
5. Plant an equal number of the positive selection and farmer selection seeds in adjacent plots, perpendicular to the slope.
6. Monitor the experiment; let the farmer group practice positive selection once more.
7. Harvest separately the two plots, record total weight and evaluate.

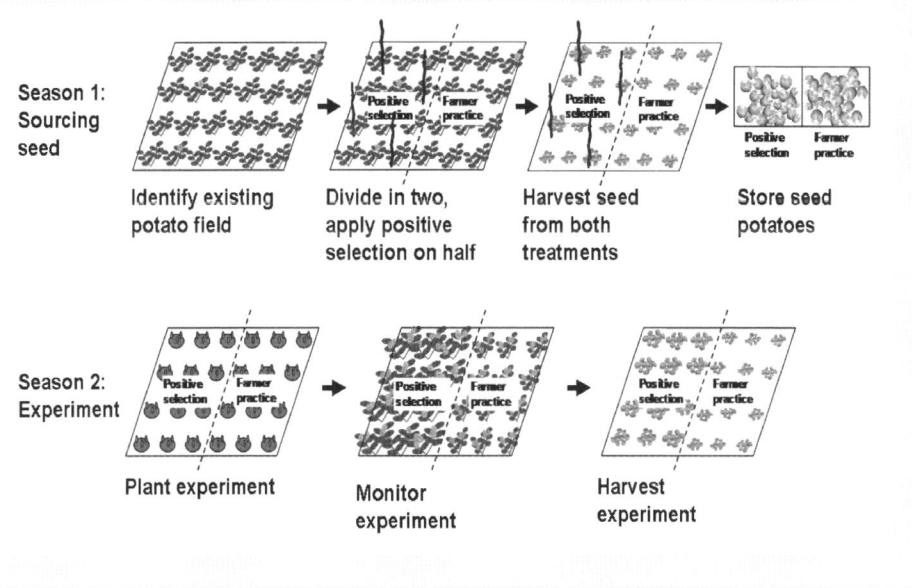

6.2.3 Positive selection programme in Kenya

A training programme on positive selection was developed, using principles of the farmer field school (FFS) method (Van der Fliert, (1993); CIP-UPWARD, (2003)). A demonstration trial formed the core of the training programme (see Box 1). The farmer group sourced seed potatoes for the trial from an existing average farmer potato field planted with a popular variety. Half of the field served to source seed potatoes using positive selection, the other half to select seed applying common farmer practice. For positive selection the farmer group pegged the best looking plants about 10 weeks after planting. Pegged plants were harvested individually but plants with few, small or misshaped tubers were rejected. Tubers measuring 25-90 mm were collected as seed potatoes. For the farmer selection treatment tubers were selected from the bulk of potatoes harvested.

The next season the demonstration experiment was planted on a field supplied by the farmer group. The field was divided into two, perpendicular to the slope. One half was planted with seed tubers obtained through positive selection, the other with seed tubers selected according to common farmer practice. All other management practices were done according to common farmer practice. Data on yield and disease incidence was collected to compare the two practices.

During the long rains of 2005 (April-July) a first pilot was executed with four farmer groups in Narok district, Kenya. The next season, during the short rains of 2005-2006 (October-January), an additional 46 farmer groups received positive selection training, spread over Narok, Nyandarua and Nakuru districts.

6.2.4 Technical results of positive selection by smallholder producers

The demonstration trials by the groups of producers showed that the proposed technology increased production by an average 34%, or 3 t/ha (Gildemacher et al., 2011) and provided substantial economic gain at little or no cash investment (Table 6.1). This corresponds with an additional benefit of 157 USD per season for an average potato producing household in Nyandarua district.

Table 6.1: Marginal net benefit of positive selection per hectare and per average household per season

Yield increase through positive selection (t/ha)	3.47
Marginal gross benefit (USD) [a][b]	367.61
Additional cost (USD) [c]	7.78
Net benefit (USD/ha)	**359.83**
Estimated benefit per adopting household (USD) [d]	**156.53**

[a] Estimated minimum farm gate price (900 Ksh / 110 kg bag)
[b] 1US$=77.12 Ksh at www.oanda.com, 01/09/2010)
[c] Casual labour = 150 Ksh / day; total 4 days / ha
[d] Avg. potato field Nyandarua 0.43 ha (Wachira et al., 2008)

6.2.5 Initial adoption of positive selection by trained farmers

The initial adoption of positive selection and farmer opinion on the technology and training approach were assessed early 2007 through focus group discussions using a checklist of questions and through individual interviews. The adoption study was carried out by a team, not engaged in the training activities. Six groups were sampled randomly in each of the three early participating districts, Narok, Nyandarua and Nakuru. Groups were sampled out of the total 48 groups that had finished their training at least 6 months prior to the survey. Technology adoption rates as a percentage of all active group members were assessed within a focus group meeting. To avoid false positives it was explained that a random selection of four adopters would be accompanied to their fields to verify the presence of marked plants. The sampled adopters were interviewed individually regarding their motives for adoption when inspecting their field, to triangulate with the answers from the group sessions.

The adoption study showed that 28% of the farmers from the sampled groups had adopted the technology of positive selection. In Nakuru, adoption figures were substantially higher (46%) than in Nyandarua (19%) and Narok (18%). Main reasons for adoption were the ability to produce quality seed at the own farm, hoping for increased yields and improving disease management. Main reason for not adopting was the drought that made that many producers farmers did not even harvest enough potatoes for replanting a crop, and had to resort to neighbours or the market to purchase seed tubers for planting the next crop. Another important reason mentioned was not having understood the technology, mainly because of absence during training sessions. The individual adopters sampled estimated that their yield increase since adopting the technology was on average 135%.

6.2.6 Discussion of the results

The yield increases estimated by the adopters were substantially higher than measured in farmer-managed trials (Gildemacher et al., 2011). Possibly, participating in the training motivated farmers to also change other crop husbandry practices, and to pay closer attention to their potato farming enterprise. The training curriculum and the demonstration trial focussed on positive selection, but also other potato crop husbandry and crop protection issues were discussed. The estimates by producers were not based on measurements and should be interpreted with care. Nevertheless the yield increase estimates confirm the usefulness of the training approach, its potential for improving productivity, and the satisfaction of the adopters with the technology of positive selection.

The initial adoption rate of 28% of the trained farmers is relatively high. The technology answers the most pressing preoccupation of potato farmers, which is seed quality (Gildemacher et al., 2009b; Gildemacher et al., 2009c). It requires no cash investment and is easy to communicate according to extension staff involved. It also fits in the existing potato seed system in which self- and neighbour supply are the most important sources of seed (Gildemacher et al., 2009a). The differences in adoption rate between the three districts are striking. The programme facilitation by the district agricultural office in Nakuru was the most effective. It was actively supervised and run as a programme fully owned by the district agricultural office. Moreover, potato production was economically most important in Nakuru. In Nyandarua prices were under pressure as producers had a reputation of producing low-quality potatoes (Wachira et al., 2008), and there were alternative cash crops because of the vicinity of the Nairobi market and favourable agro-ecology for horticultural crops. In Narok potato was a cash crop of secondary importance compared to wheat and livestock.

6.2.7 Scaling-up

Since the first pilot training in 2005-2006 and follow-up efforts in 2007 the positive selection initiative saw many spin-off and add-on activities. The training manual was published in English (Gildemacher et al., 2007) and consecutively in French, Portuguese, Spanish, Amharic, Swahili, Kinyarwanda and Runyankore/Rikiga through different development projects in collaboration with the International Potato Center (CIP). In Kenya the training effort has been scaled-up, continuing the partnership with the Ministry of Agriculture (MoA). Training programmes copying the positive selection experience from Kenya have been initiated by CIP in collaboration with local partners in Uganda, Ethiopia, Rwanda, Malawi, Mozambique and Angola and in adapted form in Ecuador and Peru. In Burundi organizations are using the training materials without CIP involvement.

CIP staff directly involved in the training programmes calculated the total number of farmers trained since 2007 based on project documentation and their knowledge of initiatives by other organisations. At the end of 2011 more than 2000 farmer groups had received positive selection training in Rwanda, Burundi, Uganda, Malawi and Kenya alone, resulting in over 40,000 farmers trained (Table 6.2). Several organizations adopted the training method for their continued efforts in seed system improvement.

Table 6.2: Estimated number of farmer groups trained on positive selection in Rwanda, Burundi, Uganda, Kenya and Malawi from 2007-2011.

	Rwanda	Burundi	Uganda	Kenya	Malawi	Total
No. groups trained with CIP	200		310	469	690	**1669**
No. groups trained without CIP	250	50	120	25	na[a]	**445**
Average group size	27	15	18	20	19	
No. known organizations involved	8	3	3	4	na	
Total no. farmers trained	**12150**	**750**	**7740**	**9880**	**13165**	**43685**

[a] na = not available
Source: CIP country representatives

6.2.8 A case of successful innovation

Considering the high adoption rate, the substantial yield increases, and the modest training cost (Box 6.2), the positive selection initiative is an example of a research initiated and coordinated process leading to successful innovation. The extensive scaling-up to major potato producing countries in Sub-Saharan Africa, South Africa excluded, is additional proof of the success of the pilot programme in Kenya.

6.3 Specific characteristics of the positive selection intervention

This section presents the characteristics of the innovation process, with special emphasis on the role of research(ers) and on how it was influenced by organizational conditions and/or how it compares with the dominant modes of thinking and operating in research establishments. The data provided mainly stem from participant observation and action

Box 6.2 Cost – benefit analysis of the positive selection training

Estimated cost (US$ a) of a farmer group training on positive selection

	Number	2006 Cost/item	2006 Total	2010 Cost/item	2010 Total
Running costs					
Allowances facilitator	12	3.74	44.86	12.97	155.60
Transport facilitator	4	1.80	7.20	2.59	10.37
Allowances supervision MoA	2	3.74	7.48	12.97	25.93
Transport supervision MoA	300	0.24	72.69	0.38	112.81
Allowances researcher	0.5	33.23	16.62	77.80	38.90
Transport research	150	0.24	36.35	0.38	56.41
Training materials	1	27.69	27.69	28.53	28.53
Training of trainer	0.33	77.54	25.59	87.14	28.76
Total running costs			238		457
Staff time					
Staff time researcher	1	41.54	41.54	52.44	52.44
Staff time MoA supervision	2	34.62	69.23	43.70	87.40
Staff time MoA facilitator	10	13.85	138.47	17.48	174.81
Total staff time			249		315
Total per group			488		772
Total per farmer			24.39		38.60

a 1 US$ = 72.22 Ksh 01-09-2006 and 1 US$ = 77.12 Ksh 01-09-2010, www.oanda.com.

The cost of training a farmer on positive selection was estimated at 24.39 and 38.60 US dollars in 2006 and 2010, respectively. Roughly 50% constituted running costs of training, such as fuel, allowances and training materials. The other 50% was the cost of staff time of research and extension staff. The average yield increase in the farmer managed demonstration trials was 34%.

The estimated benefit per household adopting the technology is 156.53 USD per season (Table 6.1). This means that in 2010 it would require a total of roughly 5 households to adopt the technology for a single season on their area under potato cultivation to 'earn back' the investment in training an entire group of 20 farmers.

research by the main researcher from mid 2004 until September 2007. In addition a 1-day participatory evaluation workshop with 24 facilitators and 4 MoA district programme managers, from Nakuru, Nyandarua and Narok district, as well as 3 participating researchers was held. This was complemented with individual interviews based on a checklist of questions with 2 MoA district programme managers, 4 facilitators, 2 farmer-facilitators and 2 researchers.

The positive selection intervention can be described by seven different features (Table 6.3). The specific choices made and the characteristics of the intervention are described hereafter. Where relevant these features are compared with the common conventions in agricultural research and development. The plausible influence of these features on the positive selection intervention is discussed.

Table 6.3: Decisions and characteristics of the positive selection initiative

	Positive selection initiative
Decision making leading to the initiative	General diagnostic studies. Coffee table discussion on alternatives for formal seed potato systems during a PRAPACE[a] meeting. Unilateral research decision making on first pilot, in an office chat.
Partnership configuration	CIP, KARI and Ministry of Agriculture district offices collaborate informally, without written agreements. Existing farmer groups are identified through the Ministry of Agriculture. Public research and extension infrastructure.
Funding of the initiative	Resource sharing to make optimal use of limited resources. Putting together funds from different projects and sources.
Freedom and flexibility	Opportunity to pursue a 'bright idea'. Freedom of engaging informally in partnership. Opportunity to adjust the initiative along the way.
Approach to validation and adaptation	Innovation focused rather than research focused. Simultaneous testing of merits technology and developing an effective training method. Training method and technology development and testing under the real circumstances of Kenyan potato farming.
Characteristics of the technology	Technically sub-optimal and low-tech manner of seed potato production. Relatively simple and adapted to smallholder reality. Technical innovation suitable for individual adoption.
Training methodology	Using elements of farmer field schools. Adaptation to available human and financial resources.

[a] Regional Potato and Sweetpotato Improvement Network for Eastern and Central Africa

6.3.1 Decision making leading to the initiative

Good practice in agricultural research and development prescribes formal and organized participation of end-users in assessing needs and opportunities, using methods such as Rapid Rural Appraisal (Chambers, 1983), Participatory Rural Appraisal (Worldbank, 1996) and Rapid Appraisal of Agricultural Knowledge Systems (Engel, 1997). In line with this, general diagnostic studies were implemented (Gildemacher et al., 2009a; Gildemacher et al., 2009b).

Klerkx and Leeuwis (2008b) stated that in addition to a comprehensive understanding of interrelations between actors it is essential to empower end-users to become decision makers in R&D. The identification of promising opportunities for improvement (Douthwaite, 2002; Van der Fliert and Braun, 2002) is specifically a moment for interaction with end-users to assess existing local innovations or indigenous knowledge.

In this case, however, the choice of positive selection emerged from a researcher-dominated perspective, during informal 'coffee table chatting' between researchers. A decision to allocate resources was taken during 'informal office chat' between CIP scientists (Box 6.3). It was agreed that improving the seed potato quality maintenance by ware potato farmers could be a worthwhile alternative seed system intervention pathway. For this, positive selection was identified as a promising technology. This was a 'top down' choice by researchers. Pretty (1994) would characterise this as passive participation by the end-users in the programme, as they had no influence on decision making. Hall (2007) called project design by researchers alone 'the worst kind of patronising behaviour'. Following Douthwaite (2002) the early decision making could be described as the moment that an individual comes up with 'a bright idea', and actually takes action to pursue the idea.

Although formal needs and opportunities were not formally assessed, the decision making was not entirely void of understanding of the local context. The decision makers were capable of assessing the situation quickly and informally. Moreover, the national researchers involved were well embedded in the potato system and had, as Klerkx and Leeuwis (2008b) rightly considered essential, a comprehensive understanding of the interrelations between actors in the system.

Box 6.3 Events and circumstances leading to the positive selection initiative

Coffee table discussion and institutional memory
CIP scientists chat over coffee during a regional potato meeting and discuss options for quality maintenance by smallholders under high disease pressure. Positive selection is mentioned as a low-tech solution that was used to initiate the seed potato multiplication system in Rwanda in the 1980s (Haverkort, 1986).

Flexible funding and freedom of decision making
Two weeks later CIP scientists sharing an office discuss what to do with $1,000 available for 'farmer based seed potato production in Sub-Saharan Africa'.

Risk taking
It was quickly agreed to test the technology of positive selection in the hands of smallholder ware potato producers, based on the flimsy evidence of one scientist having tried positive selection once himself, and the other having chatted about it over coffee.

Established partnerships and existing structures
A pilot with four farmer groups in Narok district was initiated in collaboration with the Ministry of Agriculture and the Kenyan Agricultural Research Institute.

Perseverance, grass roots involvement and enthusiasm
Two out of four pilot groups failed to conclude the trial, but the results in the other two groups, and the enthusiasm of extension staff and the two groups made the CIP scientists decide to continue, and scale up the initiative with funds from different formal projects on a substantial scale, thus taking further risks.

6.3.2 Project partnership configuration

An important feature of the positive selection initiative was the fairly traditional partnership between international research, national research, public extension, and farmer groups. The partnership consisted of CIP, Kenya Agricultural Research Institute (KARI) and district offices of the public extension service of the MoA. This very much resembles the conventional partnership associated with the much criticized linear model (Arnold and Bell, 2001). In response to widespread dissatisfaction with the linear model, there is a tendency in agricultural research and development projects to diversify partnerships. In addition, there is substantial criticism on the functioning of public agricultural services, and their limited ability to respond to end-users' demands, leading to a process of privatization of agricultural services (Anderson and Van Crowder, 2000; Heemskerk et al., 2008). Involving private entrepreneurs in agricultural research and development initiatives has, for good reasons, become more common. The contribution of public research and extension at times seems to have become redundant. In this case, however, these conventional public organizations formed the partnership.

Scientists from CIP provided seed potato experience, didactic expertise and limited financial resources. KARI provided strong understanding of the local context and the local network of actors. MoA provided locally present personnel with basic training in agriculture and local intelligence on farmer self-help groups potentially receptive to the positive selection training. The initiative collaborated directly with the district level administration of MoA. The district level assured a more than local effect of the initiative, but also allowed hands-on management by the district level extension staff. Interaction between the district management of MoA and researchers was informal. At district level collaboration was easy allowing an enabling environment in which mistakes could be made, learning was possible and practical results could be obtained, with very limited protocol or administration. Only once first results became apparent, a closer relationship between the initiative management and MoA headquarters in Nairobi was sought to improve embedding of the initiative within MoA, leading to joint assessment of opportunities for further collaboration in other potato districts.

6.3.3 Funding

The initiative was not separately funded, but financed by resources pulled together from different formal potato projects of the CIP Sub-Saharan Africa project portfolio, supplemented with in kind counterpart contributions from MoA and KARI and some ad-hoc local donor support. The first pilot was financed through US$1,000 available from core CIP funding (Box 6.3). The first results were promising enough to allocate funds from formal projects to positive selection activities, as they seemed to be relevant for smallholder potato producers.

Each partner co-funded the project to its possibilities allowing for moderation of costs and building ownership within the different partner organizations. KARI and MoA mainly contributed through staff time, while farmer groups contributed time, land and inputs. Funding was channeled to MoA district offices in such a manner that the field agents did not notice a difference with their normal recurrent budget. This made it easy for districts to continue supporting the efforts when the limited funding from CIP ran out.

6.3.4 Freedom and flexibility

A striking feature of the positive selection initiative was the freedom of manoeuvring. The organizations involved allowed their staff to invest the time and resources needed for the first pilot testing and to engage in a partnership without formalities. This freedom of manoeuvring resulted from a combination of different factors. Firstly, the organizations involved gave their staff a fair amount of liberty in decision making. Secondly, the staff involved did not hesitate to take risks, while keeping superiors abreast and informed. Thirdly, the funding used was modest, not attracting any major scrutiny. Fourthly, relations and trust existed between the organizations. Much as a result of this informal set-up, there was room for making errors, learning and taking risks. This resembles the kind of 'science culture' that Byerlee and Alex (1994) advocate.

In hindsight, the opportunity to pursue the 'bright idea' emerging around the coffee table with little but 'free, non-allocated' money (Box 6.3) appeared a critical success factor. It allowed the set-up of the first pilot with four farmer groups and few extension staff in Narok district without lengthy justifications and proof of concept. From there the initiative could evolve over time in an adaptive manner. By 2007, it had intervened with different levels of intensity in Narok, Nakuru, Nyandarua, Bomet, Buret, Meru Central and Meru North districts of Kenya.

From 2004-2007 the positive selection initiative was never a formal project in the administrative sense. It existed fairly informally in the space between projects. Although reporting on results was required, no clear milestones, activity plans and deliverables were formulated, and hence the liberty in design and adaptation of the intervention was large. There were no strict guidelines on spending the funds, which allowed for reactive use of the limited resources, there where they were most needed. Areas, facilitators and farmer groups could be selected by MoA staff.

6.3.5 Type of technology

Positive selection is a technically sub-optimal method for seed potato multiplication. Other methods, based on disease-free stock, rapid multiplication under highly controlled conditions and consecutive professional multiplication can supply ware producers healthier seed with higher yield potential. Therefore, these improved technologies have become the mainstream solutions for seed potato multiplication systems around the world. An indicator for this is that at the moment the initiative started, only four references to positive selection could be found in literature (Bokx and Want, 1987; Bryan, 1983; Salazar, 1996; Struik and Wiersema, 1999), and not a single research paper related to the merits of the technology could be traced. Clearly the technology of positive selection was considered obsolete.

In the positive selection initiative, the decision was made to invest considerable efforts in this sub-optimal technology, although technically effective manners to safeguard the best quality seed potato were at hand. Moreover, promoting positive selection meant accepting the role of ware potato farmers in seed potato multiplication. This required stopping to strive for the ideal situation in which ware potato producers purchase seed potatoes from specialized seed potato growers. This may not sound radical to an outsider, but to the potato research and development community this meant accepting an alternative pathway of seed system improvement that had always been considered a bad option. This mindset is illustrated by the following remark from a senior potato researcher: "I do not understand why you are condemning these people to poverty".

Positive selection is simple. It is fairly easy to understand that healthy looking, vigorous plants provide healthier seed tubers. Pegging such plants and harvesting them separately requires no other investment than time and sticks (Gildemacher et al., 2011). Moreover, positive selection fits within the current system: no additional organizational or institutional change is needed for an individual producer to benefit from adopting the technology.

6.3.6 Approach to validation and adaptation

The positive selection initiative focused from the start on innovation, rather than seeking research outcome alone, having the following objectives:

1) Testing the value of positive selection for smallholder potato producers;
2) Developing a cost-effective method of training smallholder potato farmers in the use of positive selection;
3) Training many potato farmers in positive selection;
4) Developing a generically applicable approach for promotion of positive selection outside Kenya.

Thus, the researchers carved a space for themselves to play wider service provision roles. Research outcome was considered an important by-product, rather than priority. The final objective was innovation, reaching beyond proving and demonstrating the merits of positive selection. The role and phasing of research was also special in that several steps of validation and adaptation were deliberately synchronized. The positive selection initiative skipped the usual steps of problem and opportunity assessment and of testing the technology under researcher control. Instead, positive selection was immediately tested under full farmer control, while simultaneously developing a method for dissemination. The initiative wanted to prove the value of positive selection in the hands of smallholder ware growers, not its technological soundness. The researchers also wanted to develop and evaluate a method for training farmers in the use of the technology as only an extensive adoption of the technology will improve the overall seed potato quality, and thus alleviate poverty. Moreover, the researchers wanted to avoid that the technology would never get beyond the stage of being 'promising' and were eager to engage with ware potato producers without delay, to assist in solving an urgent pressing problem.

From the first moment of piloting, over-nurturing was avoided to allow for both the technology and the training approach to be put to the test of reality. The technology was immediately introduced under real potato producer conditions, without giving in to the temptation to provide fertilizer, optimize soil management or disease management. The same approach was adopted regarding the coordination of the training programme by the public extension system. The MoA in the districts was provided with resources which followed the same procedures as their own recurrent funding. Supervision of efforts in the districts was left to district coordinators. Resources were stretched to the limit to achieve maximum volume of activities, much resembling the reality of the district agricultural offices. Also allowances paid were the same as set by MoA. Moreover, supervision and support by researchers were limited and much responsibility was given to MoA district offices. The partnership was such that when CIP would withdraw, two organizations with the right mandate to continue would remain, and could continue in exactly the same manner using their own recurrent budget, without the facilitators or the potato producers noticing.

It was attempted to meet both the scientist and the farmer need for experimentation simultaneously. The set-up of the trials was such that it resembled most the manner in which a farmer would experiment without scientists being involved (Bentley, 1994). The demonstration experiment of a farmer group had no replications, a single treatment was tested against current farmer practice (Box 6.1). All other management practices were decided on by the farmer group rather than by the scientists involved. The only recommendations by the scientists were to plant the same number of tubers for both treatments, in two plots alongside each other, running perpendicular to the slope and to manage both treatments the same. It was more important to show that the technology was functional in a large variety of situations than aiming for a few well executed, replicated trials.

6.3.7 Characteristics of the technology

The technology can be characterized by its ease of adoption. It requires no cash investment, nor collective action, nor system change to effectively apply and benefit from. It can be understood intuitively and even half-perfect execution will give a positive result.

6.3.8 Training methodology

The most important features of the training methodology developed were the focus on a single technology, and the fairly prescriptive training outline.

Producers were offered a single technology to solve a single important problem. In the training curriculum the alternative of buying high-quality seed potatoes from seed growers was discussed, and some high-quality seed potatoes were provided to each farmer group to demonstrate their value in obtaining higher yields. Currently in Kenya the programme compares in the same demonstration trial seed from farmer practice, positive selection and certified seed, thus providing the producer with two options for solving one problem.

The characteristics contradict with some principles of the farmer field school (FFS), in which farmers are offered a more holistic package of information, participants largely determine the agenda and the facilitator mainly stimulates learning among the participants (CIP-UPWARD, 2003). The positive selection training outline was prescribed to the trainer in rather detailed manual. The training consisted of eight sessions, spread over one and a half season, less intensive than recommended for FFS (Van der Fliert, 1993). Each training session consisted of several steps that the facilitator could follow. The training sessions had a fairly fixed content, with much less room for participants to determine the topics of the sessions than usual in FFSs. The choice for a focussed, fixed curriculum was made to avoid the need for lengthy training of trainers and to facilitate the adoption of the training methodology by trainers and facilitators with limited didactic skills.

Still, the positive selection training method used several FFS elements, including using the field as the classroom, season-long training, learning by doing through a farmer managed experiment, and discovering in the field.

6.4 Analysis and discussion

The presented experiences of positive selection in Kenya provide the setting for reflections on the role of agricultural research in innovation processes and the circumstances under which researchers can optimally perform this role.

6.4.1 Role of the researchers

The approach taken in this initiative was perhaps unconventional in some respects, but rather conventional in others. Few of the newly advocated roles for research were very prominently present: the researchers did not assess demand (they merely assumed it), did not invest much in participatory processes and strategies to ensure accountability to users (they felt they had the right expertise and mandate to implement their ideas, using mainstream extension methods), did not create an enabling social environment (focussing only on technology change) and did not actively stimulate interaction in public and private stakeholders' networks. While the partnership was indeed important, the partners chosen were conventional public organizations, and little was invested in facilitating interaction among them. The fact that these roles were not fulfilled here does not refute their importance, but it shows, based on practical evidence, that important contributions can be made to innovation by well informed and embedded conventional research-extension coalitions.

The more unconventional dimensions of the roles played by the researchers seem to relate to: (a) the responsibilities they took in training and (b) the goal and phasing of their activities.

6.4.2 Research responsibilities in the training

In the case of this intervention the researchers went beyond the conventional task of identifying entry points for change and leading the process of testing and adapting new practices and approaches, but assumed several responsibilities in training. Especially in developing and testing the training methodology researchers were highly involved, even more so than the MoA agricultural extension department.

The entire initiative was research initiated and research coordinated. Researchers developed the training programme through personally training the first four farmer groups in the very first pilot and trained all facilitators throughout. Subsequently the researchers remained actively involved by continuing to attend and sometimes facilitate selected meetings of farmer groups. This engagement enabled direct feedback, understanding of difficulties of facilitators, and close informal relationships between project managers, farmers, facilitators and coordinators that allowed for hands-on, informal and effective monitoring and project management. Over time the field presence during the training sessions by the researchers was reduced and attention was put on initiating the same activity in other districts.

A final important responsibility that researchers assumed regarding the training was upscaling. They actively promoted the technology internally in their own organizations and externally with development organizations. They secured funding for implementing the training programme in additional locations in and outside Kenya, and succeeded to embed the use of the training method in the activities of organizations with the routine mandate to train agricultural producers.

6.4.3 Goal and phasing of activities

The initiative focused on innovation, rather than research or technology development. This shifted the focus to training, but also to an approach deviating from the usual order of research and extension. The conventional 'linear' model of innovation (Godin, 2006) consists of a phase of fundamental research, then applied research, after which results are published and brought to the attention of 'development partners' with a mandate for agricultural extension. Douthwaite (2002) describes the same sequence of events in agricultural technology development, while acknowledging that several cycles of collaborative learning and adapting with farmers are required in the field before a technology gets adopted by more end-users. In this case, however, neither of these models was adhered to. Instead, all these activities were carried out simultaneously. The usual step of testing the technology under researcher control was omitted, and the technology was immediately tested under full farmer control. Moreover, these activities were immediately integrated in extension activities and in curriculum design for further communication and upscaling. This was all part of a deliberate strategy. The emphasis was not on testing the technology as such, but on its robustness in the hands of farmers across a wide range of situations. The philosophy was that, eventually, it is only worth investing in massive dissemination of a technology if yield benefits are robust and substantial, and can thus be demonstrated convincingly to farmers in non-replicated trials.

The choice to test the technology immediately within 'real life' conditions not only deviates from the linear and/or iterative conceptualizations of the role of research in innovation, but also from the more recent idea that innovation temporarily requires an experimental space (or 'niche') that can be shielded off and protected from the regular environment, to allow learning and improving the innovation design (Geels and Raven, 2006; Kemp et al., 2001; Schot and Geels, 2008). Like the linear model, this so-called 'strategic niche management' perspective legitimizes the separation between a research and design phase under artificial conditions, and a later phase of dissemination and upscaling. In this study we observed that such a separation is not always necessary, and that both activities can be fruitfully combined. Through the deliberate exposure to, and the consideration of the prevalent circumstances of both producers and public organizations, a non-replicable pilot success was avoided. This did not only lead to developing and demonstrating a sound technology, but also to developing and using a time and cost effective method for dissemination. This experience supports the view expressed by Hommels et al. (2007) that niches should not be over-protected and that learning and adaptation may best occur under real-life conditions rather than in an artificial experimental setting. Moreover, it supports the idea that dissemination should not be considered as an end-of-pipe phenomenon, but as something that already starts from the early development stages (Leeuwis and Aarts, 2011).

6.4.4 Enabling or hindering conditions within the research organization

One could argue that since the initiative evolved from an international research organization, there is apparently sufficient space within such institutes to effectively innovate agriculture. This is partly true, as research organizations have considerably advanced in their thinking about innovation, and have also adopted new ideas and discourses about the role of researchers (Anandajayasekeram, 2011). However, jumping to such a conclusion would ignore that the positive selection initiative created its own space where it was safe-guarded from the dominant ways of doing things within the research organization. The technology promoted went against the mainstream policy and strategy of the seed potato research establishment. An important condition enabling the researchers to go ahead with the initiative was that it was never a formal project and required minimal resources. Moreover it was executed within a regional programme of the international research organisation, where more freedom of manoeuvring was possible than at headquarters. The regional office was characterised by having few staff with a high decision making freedom. This created space for researchers to respond to emerging opportunities and demands with considerable flexibility and informality. The room of manoeuvring that characterizes the positive selection initiative is not standard in donor funded agricultural research and development interventions. Usually projects promise deliverables according to a pre-designed time frame, with specific milestones, to reassure their funders that they get the results they aspire. Once getting into the phase of implementation, the monitoring, evaluation and accountability protocols of both implementing and funding organizations do not allow for important deviations from plans. We think it is unlikely that this project would have been accepted and funded if its purpose and process design had been spelled out in a formal project proposal.

An enabling condition to this initiative resides in combining available skills and expertise. The combination of national researchers with much locally embedded knowledge of the potato sector Kenya and international scientists with technical potato expertise and experience in participatory research and development proved effective. The national research programme did not only provide expertise in technical issues, but also provided the required practical understanding of the Kenyan potato chain functioning. The international researchers also had experience in collaborating with deficient public organizations. This allowed for avoiding many pitfalls of misjudging opportunities and underestimating the complexity of farmer practice, and made up for the lack of a formal assessment of needs and opportunity.

We conclude that while there was an enabling environment for piloting, this did not result from any deliberate policy to facilitate innovation. Rather, it resulted from the willingness of individuals in the participating organizations to take risks by going against the conventional wisdom of the research organizations, to collaborate with little resources and ignore where necessary protocol. The leeway that partners had was mainly caused by the modest investment and associated little scrutiny from CIP, KARI and MoA.

6.4.5 Enabling or hindering external conditions

In current literature on innovation (systems), the need to combine and align technology change with institutional change is emphasised, for example in creating supportive policies, conducive credit facilities, appropriate marketing arrangements and better extension (Dormon et al., 2007; Hall, 2005; KIT and CFC, 2011; Klerkx et al., 2009; Leeuwis, 2004; Leeuwis and van den Ban, 2004). This emphasis arises from the idea that institutional change ('orgware') is an integral component of innovation, but can –in practice- also be informed by pressure within agricultural research and extension projects to deliver success. In this case, however, no attention was paid to building an enabling social environment for technology use. To the contrary, in contrast to most earlier seed system interventions aiming to build an entire specialized seed grower system, the positive selection intervention accepted existing institutions. The researchers involved invested in understanding the existing institutions and focused, with success, on incremental innovation, within the limits and opportunities of the existing system.

Several external enabling factors, beyond the influence of research, can be identified. The existence of a dynamic marketing system and the trend of increasing consumption (Gildemacher et al., 2009b) provided a market for increased surpluses produced by farmers, while the Kenyan public extension infrastructure and widespread farmer self-help groups facilitated the implementation of farmer training.

Furthermore the innovation was not complex and fitted within the existing potato production system. It is therefore surprising that this technology was not widely used already. Adoption of the technology is possible without any major farming system, organizational or institutional change. Thus, it was not a 'system innovation' (Leeuwis and Aarts, 2011; Schot and Geels, 2008) requiring new forms of coordination among various stakeholders, but a simple practice that individuals can adopt. Only knowledge of the technology is required, which can be provided through straightforward training by agents with a basic level of training.

The case of positive selection points to the fact innovation is sometimes possible within the existing institutional reality, without institutional change. Considering the complexity of realising institutional change, it is worthwhile to specifically search for incremental innovation that can, as in our case, provide positive change within the existing system.

6.4.6 Relevance of the case to other innovation trajectories

To what extent are the features of the positive selection initiative relevant to other agricultural innovation trajectories? The positive selection innovation was simple and technical. It required software (knowledge change of producers, extension staff and researchers) and hardware (the technology development itself) changes, but hardly any 'orgware' adaptations.

Arguably, many other innovations in agriculture (e.g. new varieties, tools, irrigation techniques) are collective in nature (Rogers, 1995, 2003) and require simultaneous change within a network of stakeholders (Leeuwis and van den Ban, 2004). In such cases, more attention will be needed for playing the new 'innovation intermediation' roles (e.g. demand articulation, facilitating interaction in networks), aimed at aligning technological change and institutional change. The more conventional roles and partnerships reflected in this case are probably insufficient in more complex innovation settings.

Still, the research conditions that enabled this initiative to start and adapt flexibly, are likely to be relevant also for more complex innovation settings. The more dimensions and stakeholders are involved, the greater the chances that conditions, insights and relations among stakeholders change over time, calling for flexibility and deviation from pre-planned goals and activities. Moreover, more complex socio-technical innovations can only really be designed, tested and adapted through co-evolution within a societal environment, and do not allow for complete testing in 'on station' research. The approach of simultaneous researcher and user experimentation under 'real life' conditions may also be relevant to more complex innovation trajectories.

6.5 Conclusion and implications for research and development

This specific case study has demonstrated that agricultural innovation does not always require institutional change. Identifying opportunities for incremental innovation within existing institutional settings is worthwhile, as they are easier to realise than system innovations. At the same time, the case suggests that researchers may take on far greater responsibilities in training and curriculum development than is usually considered appropriate.

The positive selection case points out subtly that the continuous search for new technological opportunities for change may lead to overlooking the obvious simple solutions. It is a clear reminder that innovation is not about developing new technology, but about adoption and adaptation of practices in new environments. Although positive selection is old and technically sub-optimal, its adaptation for and adoption by smallholder farmers is a radical innovation of potato seed systems. It implies a shift of focus from specialised seed producers as the unique pathway to seed quality improvement to a system in which also quality maintenance by smallholder ware potato producers plays a role. There might be more 'old' and/or unfashionable technologies and practices that are worth (re)considering in African farming systems. In view of the rapid pace of technology development and turnover of ideas in science, which is often out of sync with the speed of innovation in practice, and the continuous handing over of agricultural production from one generation of farmers (and researchers and extensionists) to the next, it is quite likely that useful ideas, originating from either science of local experience, become more or less forgotten. Thus, both the anticipated future and history may be a good inspiration for generating new ideas that are worth exploring.

It is necessary to have an enabling environment for 'bright ideas', to be identified and pursued, even if they go against the dominant solution strategies at the time. In the case of positive selection a small fund without fixed deliverables attached, the opportunity to forge a partnership without administrative formalities and the opportunity to consecrate human resources to the initiative were essential. Researchers should create space for manoeuvre that allows them to pursue options that are not considered relevant, or in this case even sub-optimal and outdated. This observation indicates that agricultural research and development organizations need to consider how similar freedom and opportunity to identify and pursue ideas can be created deliberately.

The positive selection initiative benefitted from flexibility in design and implementation. There was ample room for manoeuvring and adjustment of the initiative based on experiences gained in the process. The idea that development and innovation can be stimulated through an orderly planned trajectory has been refuted and criticised by many scholars in development sociology and innovation studies (Crehan and Von Oppen, 1988; Leeuwis, 2004; Long and Van Der Ploeg, 1989; Loorbach, 2007; Schot and Geels, 2008). Rather than focussing mainly on accountability against fixed milestones and indicators, more appreciation for problem solving performance and quality of the process of searching for solutions and learning may assist in enhancing the contribution of research to agricultural innovation.

If one accepts that the objective of research is to contribute to agricultural innovation, a reconsideration of the role and position of researcher may be opportune. In our case the role of research went much beyond proving theory; in partnership with development organizations responsibility was assumed to make the new insights and technology work in practice. This case shows that commitment can be found in international and national research organizations to remain involved in agricultural innovation initiatives beyond the point of data collection and analysis, and engage actively with other partners in all stages of innovation, from pursuing a first bright idea, to developing effective local adaptation and communication strategies and advocacy for mainstream acceptance. The current perception of the mandate of agricultural research does not always allow applied research to actively partner in development processes. This partnership, however, allows for research to remain involved, be informed and inspired by agricultural producers, traders and advisory service providers. It is through this interaction that new ideas and opportunities can emerge and new process of innovation can be sparked. Actively supporting the continued immersion of agricultural research in development practice will contribute to both the quality and client orientation of research, and the same time add quality to the development process through the contribution by researchers.

References

Adjei-Nsiah, S., C. Leeuwis, K. E. Giller, O. Sakyi-Dawson, J. Cobbina, T. W. Kuyper, M. Abekoe and W. Van Der Werf 2004. Land tenure and differential soil fertility management practices among native and migrant farmers in Wenchi, Ghana: Implications for interdisciplinary action research. NJAS - Wageningen Journal of Life Sciences 52: 331-348.

Anandajayasekeram, P 2011. The role of agricultural r&d within the agricultural innovation systems framework. In The role of agricultural r&d within the agricultural innovation systems framework, Agricultural R&D—Investing in Africa's Future: Analyzing Trends, Challenges, and Opportunities, 1-33. Accra, Ghana: ASTI, FARA.

Anderson, J. and L. Van Crowder 2000. The present and future of public sector extension in Africa: contracting out or contracting in? Public administration and development 20: 373-384.

Arnold, E and M Bell 2001. Some new ideas about research for development. In In: Partnership at the Leading Edge: A Danish Vision for Knowledge, Research and Development ed. DANIDA, 279-316. Copenhagen: Danish Ministry of Foreign Affairs.

Biggs, S. 2007. Building on the positive: An actor innovation systems approach to finding and promoting pro poor natural resources institutional and technical innovations. International Journal of Agricultural Resources, Governance and Ecology 6: 144-164.

Bokx, JA de and JPH van der Want 1987. Viruses of potatoes and seed-potato production. Wageningen: Pudoc.

Bryan, JE 1983. On-farm seed improvement by the potato seed plot technique. In On-farm seed improvement by the potato seed plot technique, 13. Lima: International Potato Center.

Chambers, R. 1983. Rural development : putting the last first. London: Longman.

CIP-UPWARD 2003. Farmer Field Schools: From IPM to Platforms for learning and Empowerment. Los Baños: CIP-UPWARD, ARD, Rockefeller Foundation.

Crehan, K. and A. Von Oppen 1988. Understandings of "development": an arena of struggle. The story of a development project in Zambia. Sociologia Ruralis 28: 113-145.

Crissman, C. C., L. M. Crissman and C. Carli 1993. Seed potato systems in Kenya: a case study. Lima: CIP.

Devaux, A., D. Horton, C. Velasco, G. Thiele, G. Lopez, T. Bernet, I. Reinoso and M. Ordinola 2009. Collective action for market chain innovation in the Andes. Food Policy 34: 31-38.

Dormon, E. N. A., C. Leeuwis, F. Y. Fiadjoe, O. Sakyi-Dawson and A. van Huis 2007. Creating space for innovation: the case of cocoa production in the Suhum-Kraboa-Coalter District of Ghana. International Journal of Agricultural Sustainability 5: 232-246. doi: 10.1080/14735903.2007.9684824

Douthwaite, B. 2002. Enabling innovation : a practical guide to understanding and fostering technological change. London: Zed Books.

Edquist, Charles and Björn Johnson 1997. Institutions and Organizations in Systems of Innovation. In Systems of innovation; technologies, institutions and organizations, ed. Charles Edquist, 41-63. London; Washington: Pinter.

Engel, P. G. H. 1997. The social organization of innovation : a focus on stakeholder interaction. Amsterdam: Royal Tropical Institute.

Engel, PGH 1995. Facilitatign Innovation: an actor oriented approach and participatory methodology to improve innovative social practice in agriculture, Wageningen Agricultural University.

Eshetu, Mulatu, O. E. Ibrahim and Bekele Etenesh 2005. Improving potato seed tuber quality and producers' livelihoods in Hararghe, Eastern Ethiopia. Journal of New Seeds 7: 31-56.

Funtowicz, S. O. and J. R. Ravetz 1993. Science for the post-normal age. Futures 25: 739-755.

Geels, F. and R. Raven 2006. Non-linearity and expectations in niche-development trajectories: Ups and downs in Dutch biogas development (1973-2003). Technology Analysis and Strategic Management 18: 375-392.

Getachew, T and A. Mela 2000. The role of SHDI in potato seed production in Ethiopia: Experience from Alemaya integrated rural development project. In The role of SHDI in potato seed production in Ethiopia: Experience from Alemaya integrated rural development project, 5th African Potato Association Conference, 109-112. Kampala, Uganda: African Potato Association.

Gildemacher, P.R., P. Maina, M. Nyongesa, P. Kinyae, W. Gebremedhin, Y. Lema, B. Damene, T. Shiferaw, R. Kakuhenzire, I. Kashaija, C. Musoke, J. Mudiope, I. Kahiu and O. Ortiz 2009a. Participatory Analysis of the Potato Knowledge and Information System in Ethiopia, Kenya and Uganda. In Innovation Africa: Enriching farmers' livelihoods, eds. P.C. Sanginga, A. Waters-Bayer, S. Kaaria, J. Njuki and C. Wettasinha, 153-167. Sterling: Earthscan.

Gildemacher, Peter, Paul Demo, Ian Barker, Wachira Kaguongo, Gebremedhin Woldegiorgis, William Wagoire, Mercy Wakahiu, Cees Leeuwis and Paul Struik 2009b. A Description of Seed Potato Systems in Kenya, Uganda and Ethiopia. American journal of potato research 86: 373-382. doi: 10.1007/s12230-009-9092-0

Gildemacher, P., W. Kaguongo, O. Ortiz, A. Tesfaye, G. Woldegiorgis, W. Wagoire, R. Kakuhenzire, P. Kinyae, M. Nyongesa, P. Struik, and C. Leeuwis. 2009. Improving Potato Production in Kenya, Uganda and Ethiopia: A System Diagnosis. Potato Research 52:173-205.

Gildemacher, Peter, Elmar Schulte Geldermann, Dinah Borus, Paul Demo, Peter Kinyae, Pauline Mundia and Paul Struik 2011. Seed Potato Quality Improvement through Positive Selection by Smallholder Farmers in Kenya. Potato Research 54: 253-266. doi: 10.1007/s11540-011-9190-5

Godin, B. 2006. The linear model of innovation - The historical construction of an analytical framework. Science Technology & Human Values 31: 639-667. doi: 10.1177/0162243906291865

Hall, A 2007. The origins and implications of using innovation systems perspectives in the design and implementation of agricultural research projects: Some personal observations. In The origins and implications of using innovation systems perspectives in the design and implementation of agricultural research projects: Some personal observations, Working Paper Series, 22. Maastricht: UNU-MERIT.

Hall, A. 2005. Capacity development for agricultural biotechnology in developing countries: An innovation systems view of what it is and how to develop it. Journal of International Development 17: 611-630.

Hall, A., G. Bockett, S. Taylor, M. V. K. Sivamohan and N. Clark 2001. Why research partnership really matter: Innovation theory, institutional arrangements and implications for developing new technology for the poor. World Development 29: 783-797.

Hall, Andy 2006. Publicprivate sector partnerships in an agricultural system of innovation: Concepts and challenges. International Journal of Technology Management & Sustainable Development 5: 3-20.

Haverkort, A. 1986. Forecasting national production improvement with the aid of a simulation model after the introduction of a seed potato production system in central Africa. Potato Research 29: 119-130.

Heemskerk, W., S. Nederlof, B. Wennink and B. Shapland 2008. Outsourcing agricultural advisory services : enhancing rural innovations in Sub-Saharan Africa. Amsterdam: Royal Tropical Institute (KIT), KIT Development, Policy and Practice.

Hirpa, Adane, Miranda Meuwissen, Agajie Tesfaye, Willemien Lommen, Alfons Oude Lansink, Admasu Tsegaye and Paul Struik 2010. Analysis of Seed Potato Systems in Ethiopia. American journal of potato research 87: 537-552. doi: 10.1007/s12230-010-9164-1

Hocdé, H, B Triomphe, M Faure and M Dulcire 2008. From participation to partnership: A different way for researchers to accompany innovation porcesses - challenges and difficulties. In Innovation Africa: Enriching farmers' livelyhoods, eds. P.C. Sanginga, A. Waters-Bayer, S. Kaaria, J. Njuki and C. Wettasinha, 89-103. Sterling: Earthscan.

Hommels, A., P. Peters and W. E. Bijker 2007. Techno therapy or nurtured niches? Technology studies and the evaluation of radical innovations. Research Policy 36: 1088-1099.

Howells, J. 2006. Intermediation and the role of intermediaries in innovation. Research Policy 35: 715-728.

Kemp, R, A Rip and J Schot 2001. Constructing transition pathsthrough the management of niches. In Path dependence and creation, eds. R Garud and P Karnoe, 269-299. Mahwah: Lawrence Earlbaum.

KIT and CFC 2011. From sorghum to shrimp: a journey through commodity projects. Amsterdam: KIT publishers.

Klerkx, L., A. Hall and C. Leeuwis 2009. Strengthening agricultural innovation capacity: Are innovation brokers the answer? International Journal of Agricultural Resources, Governance and Ecology 8: 409-438.

Klerkx, L. and C. Leeuwis 2008a. Balancing multiple interests: Embedding innovation intermediation in the agricultural knowledge infrastructure. Technovation 28: 364-378. doi: 10.1016/j.technovation.2007.05.005

Klerkx, Laurens and Cees Leeuwis 2008b. Matching demand and supply in the agricultural knowledge infrastructure: Experiences with innovation intermediaries. Food Policy 33: 260-276.

Kristjanson, P., R. S. Reid, N. Dickson, W. C. Clark, D. Romney, R. Puskur, S. MacMillan and D. Grace 2009. Linking international agricultural research knowledge with action for sustainable development. Proceedings of the National Academy of Sciences of the United States of America 106: 5047-5052.

Leeuwis, C. and A. van den Ban 2004. Communication for rural innovation : rethinking agricultural extension. Oxford: Blackwell Science.

Leeuwis, Cees 2004. Fields of conflict and castles in the air. Some thoughts and observations on the role of communication in public sphere innovation processes. The Journal of Agricultural Education and Extension 10: 63 - 76.

Leeuwis, Cees and Noelle Aarts 2011. Rethinking Communication in Innovation Processes: Creating Space for Change in Complex Systems. The Journal of Agricultural Education and Extension 17: 21-36. doi: 10.1080/1389224x.2011.536344

Long, N. and J. D. Van Der Ploeg 1989. Demythologizing planned intervention: an actor perspective. Sociologia Ruralis 29: 226-249.

Loorbach, D 2007. Transition Management: New mode of governance for sustainable development, Erasmus University.

Lundvall, B.A. 1992. National systems of innovation: towards a theory of innovation and interactive learning. London: Pinter.

Lutaladio, N, O Ortiz, A Haverkort and D Caldiz 2009. Sustainable Potato Production; guidelines for Developing Countries. Rome: FAO.

Muthomi, JW, JN Nyaga, FM Olubayu, JH Nderitu, JN Kabira, SM Kiretai, JA Auro and M Wakahiu 2009. Incidence of aphid-transmitted viruses in farmer-based seed potato production in Kenya. Asian Journal of Plant Sciences 8: 166-171.

Pretty, Jules N. 1994. Alternative Systems of Inquiry for a Sustainable Agriculture. IDS Bulletin 25: 37-49. doi: 10.1111/j.1759-5436.1994.mp25002004.x

Rajalahti, Riikka, Willem Janssen and Elijah Pehu 2008. Agricultural innovation systems: from diagnostics toward operational practice. Washington: World Bank.

Rogers, E 1995, 2003. Diffusion of innovations. New York: Free Press.

Salazar, LF 1996. Potato viruses and their control. Lima: International Potato Center (CIP).

Schot, J. and F. W. Geels 2008. Strategic niche management and sustainable innovation journeys: Theory, findings, research agenda, and policy. Technology Analysis and Strategic Management 20: 537-554.

Schut, M., A. van Paassen, C. Leeuwis, S. Bos, W. Leonardo and A. Lerner 2011. Space for innovation for sustainable community-based biofuel production and use: Lessons learned for policy from Nhambita community, Mozambique. Energy Policy 39: 5116-5128.

Smits, R. 2000. Innovatie in de universiteit. In Innovatie in de universiteit, Inaugurele rede Universiteit Utrecht. Utrecht: Universiteit Utrecht.

Spielman, D. J., J. Ekboir and K. Davis 2009. The art and science of innovation systems inquiry: Applications to Sub-Saharan African agriculture. Technology in Society 31: 399-405.

Struik, P. C. and S. G. Wiersema 1999. Seed potato technology. Wageningen: Wageningen Press.

Tindimubona, S., R. Kakuhenzire, J.J. Hakiza, W.W. Wagoire and J. . Beinamaryo 2000. Informal production and dissemination of quality seed potato in Uganda. In Informal production and dissemination of quality seed potato in Uganda, 5th African Potato Association Conference, 99-104. Kampala, Uganda: African Potato Association.

Van der Fliert, E and A.R. Braun 2002. Conceptualizing integrative, farmer participatory research for sustainable agriculture: From opportunities to impact Agriculture and Human Values 19: 25-38.

Van der Fliert, E. 1993. Integrated Pest Management: Farmer Field Schools Generate Sustainable Practices. A Case Study in Central Java Evaluating IPM Training. Wageningen Agricultural University papers 93-3, Agricultural University Wageningen.

Wachira, K, P. Gildemacher, P. Demo, W. Wagoire, P. Kinyae, J. Andrade, K. Fuglie and G. Thiele 2008. Farmer practices and adoption of improved potato varieties in Kenya and Uganda. In Farmer practices and adoption of improved potato varieties in Kenya and Uganda, Social Sciences Working paper, 85. Lima.

Woolthuis, R. K., M. Lankhuizen and V. Gilsing 2005. A system failure framework for innovation policy design. Technovation 25: 609-619.

Worldbank 2007. Enhancing agricultural innovation; how to go beyond the strengthening of research systems. Washington: World Bank.

Worldbank 1996. The Worldbank participation sourcebook Washington DC: World Bank.

7 General discussion

7.1 Introduction

The research trajectory presented in this thesis started with the observation that an intensification of the potato production system in Eastern Africa is needed to meet the growing demands from the market and with the assumption that this provides opportunities for smallholder potato producers. This observation and this assumption were translated into the overarching research question:

'How can smallholder potato productivity and profitability be increased in Kenya, Uganda and Ethiopia?'

To answer this overarching research question a number of research efforts followed, with the results of each component determining the choices made for the next stage of research. As a first step, a potato production system diagnosis was made to identify potato production system constraints and entry points for innovation (Chapter 2). Seed potato quality improvement was identified as one of the priority entry points in the three countries. To identify innovative pathways of seed potato system improvement, a more in-depth analysis of the seed potato system in the three countries was carried out (Chapter 3). This in-depth analysis identified seed potato quality management by ware potato producers as a particular option for change. From different techniques that could contribute to improved seed potato quality management positive selection was chosen for further research. Through action research with groups of potato producers in Kenya positive selection was shown to offer important benefits to smallholder producers (Chapters 4 and 6). Additional questions emerged with regard to the explanation behind the yield increases obtained by applying positive selection in farmer fields. Replicated trials in a number of Kenyan agro-ecologies were implemented to investigate the supposition that a reduction in virus incidence could explain the yield effects of positive selection (Chapter 5). Positive selection showed to be a promising technology for smallholder ware potato producers. The full potential of positive selection for seed potato quality improvement can, however, only be realized through a widespread popularization of the technology among ware potato producers. This requires effective training and communication approaches. An assessment of the research and training approach was made to learn lessons for scaling-up (Chapter 6). Chapter 6 paid specific attention to the role of agricultural research in the process of innovation, using the positive selection research trajectory as a learning case.

In this general discussion the findings from the different components of the research trajectory are brought together, linked to broader debates and synthesized. Specifically the consequences of the research findings for efforts to improve potato systems in East Africa, and more generally in Sub-Sahara Africa, are discussed. In addition, suggestions are made for the direction of further applied and more fundamental research to contribute to seed potato system innovation. Finally, the general discussion will elaborate further on the insights that can be derived from the entire research trajectory with regard to the contribution of agricultural research(ers) to agricultural innovation.

7.2 Potato system diagnosis

The potato system diagnosis presented in Chapter 2 combined quantitative and qualitative data collection methods. The diagnosis assisted in identifying priority areas for intervention to improve the sector, thus contributing to decision making with regard to options for deliberate innovation efforts. Seed potato quality management, bacterial wilt (*Ralstonia solanacearum*) control, late blight (*Phytophthora infestans*) control and soil fertility management were identified as key technical intervention topics for innovation. Based on the results of the study further research work was initiated in the field of seed system improvement and integrated management of late blight (Gildemacher et al., 2007c; Gildemacher et al., 2007d; Kakuhenzire et al., 2007a; Kakuhenzire et al., 2007b; Nyongesa et al., 2007; Nyongesa et al., 2005) and potato bacterial wilt (Kwambai et al., 2011; Wakahiu et al., 2007). This thesis only reports on further research in seed potato quality improvement.

Furthermore it was concluded that the potato marketing system in the three countries required attention. An important constraint for potato producers to invest in intensification of their production is the insecurity of the marketing of their produce. At the same time wholesalers and processors have difficulties obtaining the quantities and quality of produce they require (Gildemacher et al., 2009c). Improving the communication and coordination between chain actors holds potential for improving the farm gate margins of producers on one side of the chain and the profit margins of processors, wholesalers and retailers and ultimately the satisfaction of consumers on the other end of the chain. Efforts to improve chain communication and collaboration were concluded to be worthwhile to pursue. Another systemic imperfection identified was the interaction between different actors with regard to knowledge. The flow of potato-related information between research, intermediary advisory services, private entrepreneurs and producers was deemed far from perfect and suggestions for improvement were made, as presented in work not incorporated in this thesis (Gildemacher et al., 2009c). The stakeholder interaction meeting organized for the system diagnosis formed a first step in improving communication between knowledge and service providers such as researchers, agricultural advisors, farmer representatives and local government.

The combination of quantitative surveys with stakeholder interaction workshops proved to be particularly useful in improving the understanding of the East African potato system. The construction of the matrix of interaction (adapted from Biggs and Matsaert (1999)) by stakeholders groups, and specifically the explicit identification of constraints in interaction between them, provided important insight in system constraints. The manner in which it was applied provided a non-hostile environment in which stakeholder groups apparently felt 'safe' to voice their concerns honestly and openly about the performance of other stakeholders. The adapted methodology used is very suitable for an accurate and quick identification of needs and opportunities in agricultural sub-systems. The system failure framework (Woolthuis et al., 2005) did provide for a pragmatic framework to combine and structure the opportunities for innovation in the potato sector, and facilitated the choice of promising entry points for innovation. The utility of the diagnostic study for the research trajectory went beyond the identification of the said entry-points.

It also provided the system overview required to assist in decision making further in the research process and weighing the consequences of research findings for the potato system practice. In addition the stakeholder interaction workshops provided the starting point of network building between actors. It went beyond being a tool to identify entry points for innovations, it was the first step in building stakeholder coalitions for innovation.

7.3 Seed potato system diagnosis

The diagnostic study of the potato system was followed by a more in-depth analysis of the seed potato system in the three countries (Chapter 3). This analysis aimed at identifying innovative approaches to improve the quality of seed potatoes used in the three countries. The study was done while keeping in mind that seed system building was not an objective as such, but was uniquely aimed at improving ware potato production. In other words, the study investigated how the quality of average seed potato planted by ware potato producers could be improved. Poverty impact was not expected from the development of seed business as such, but from the improved overall quality of seed potatoes used by smallholder ware potato growers.

7.3.1 Potato diseases

The seed potato system diagnosis confirmed the observations made in the general study presented in Chapter 2 that the current situation of the seed potato system is contributing to low potato yields, and that it deserved attention in order to improve smallholder potato productivity and profitability. Potato viruses were found to be abundant in potatoes sold as seed in rural markets in Kenya. This confirms the non-documented field observations of abundant symptoms in Kenyan potato fields. The infection incidences obtained in the study can, however, not be directly translated into field incidences. Only seed potatoes from a single source, rural open markets were assessed, while potato producers in Kenya source seed potatoes predominantly from neighbours. Furthermore Potato Virus M (PVM) was not included in the testing, whereas it is known that this virus is abundant in seed tubers in Kenya (Muthomi et al., 2009). No other systematic assessment of virus incidences in field potato crops has been conducted in Kenya, as was for example done in Plateau State, Nigeria (Miha et al., 1993). Muthomi et al. (2009) sampled only from seed multipliers in two districts in Kenya, and recorded high levels of Potato Virus S (PVS), Potato Leaf Roll Virus (PLRV) and Potato Virus M (PVM), and lower levels of Potato Virus Y (PVY) and Potato Virus X (PVX). Seed quality of specialized seed potato multipliers is, however, much like seed offered on rural markets as sampled in our study presented in Chapter 3, a misleading proxy for the average quality of seed potatoes planted in Kenya, as only a small proportion of Kenyan potato farmers, who replace their seed, do so by sourcing from such multipliers. Were et al. (2003) found field infections of PLRV across Kenya rating between 1.5% and 28.9%, which is much lower than the incidences found in our sample from rural markets. They did, however, first do a visual scoring, which was then confirmed by testing a leaf sample of a visually infected plants through DAS-ELISA, a method that would overlook symptomless infected plants.

Bacterial wilt was shown to be endemic; it could be detected in 74% of the sampled fields. The incidence was, however, fairly low with 1.1% on average (see Chapter 2). Kwambai et al. (2011) found similar high prevalence of the disease, but reported incidences as high as 7% on average in Trans Nzoya district. The expression of bacterial wilt is very erratic, and depends on many factors, including soil type, humidity, temperature, crop stage and variety (CIP, 1996; Priou et al., 2009), which makes a field assessment of infection incidences complicated.

7.3.2 Seed potato stock replacement in East Africa

The study of seed potato systems in the three countries showed that under the current farmer practice the total seed stock getting replaced by seed potatoes from outside the farm is extremely low, with 7%, 4% and 15% in Kenya, Uganda and Ethiopia, respectively. Several reasons for this low replacement rate were suggested. In the first place, there is a limited awareness about seed potato degeneration in general and potato viruses and bacterial wilt in particular, amongst producers and agricultural advisors alike. Without this awareness the incentive for smallholder producers to replace (part of) their seed stock from a reliable source is limited. Secondly, regular (partial) seed potato stock replacement requires a substantial cash investment, whereas seed potato selection from the own harvest does not require any cash. This is an important consideration for generally cash-short smallholder farmers, for whom other livelihood needs quickly take priority over saving resources for an optional investment in next seasons' crop. Adding to this are the yield and market insecurities associated with rainfed agriculture which are an important disincentive for investments in crop intensification (Shiferaw et al., 2009; Wani et al., 2009). The East African potato markets are prone to inevitably low farm gate prices during glut periods when most producers are harvesting, as there is limited to no ware potato storage capacity. In general, less endowed smallholder farmers are more prone to reducing their risks by minimizing cash investments than to investing for profit maximization. All in all, the system analysis makes plausible that there are circumstances under which it makes economic sense for smallholder potato producers to restrain from high investments in seed potato replacement. Finally, would farmers decide they are in a position to invest in high-quality seed potatoes, they will have to make an effort to find the desired variety at the right time, as seed potatoes of reliable high quality are a scarce commodity in all three countries featuring in the study.

7.3.3 Seed potato quality management by ware potato producers

In drawing conclusions from the study for seed potato system improvement, care has to be taken not to consider potato producers as a homogeneous group. Ultimately the decision to invest or not in seed potato renewal depends on variables such as market security, expected farm gate price, crop loss risks, alternative cash needs and investment opportunities, the balance between food security and income objectives of the smallholder farm enterprise and expected degeneration rates of the specific variety - agro-ecology - crop husbandry combination. A producer with reliable market access and favourable farm gate prices and access to financing could opt for profit maximization, and renew the seed potato stock regularly. A producer with poor market security and a less favourable cash position would be wise to opt for a lower renewal frequency, and rely longer on self-supply from the own harvest.

Considering the diversity in needs, a diversity of options should be offered to producers. In response to this a dual seed system development strategy is recommended. On the one hand efforts are needed to improve the availability of affordable high-quality seed potatoes through a commercial system of specialized commercial seed multipliers. At the same time, however, it must be recognized that the current common practice of producers to select seeds from their own harvest will remain an important feature of the potato production system in East Africa. The study shows that ware potato producers do this under sub-optimal conditions. There is a world to win through investigating, optimizing and promoting simple measures to improve the quality of the self-produced seed. Surprisingly, improving quality management by ware potato producers has until now hardly been considered as an important entry point for seed system innovation. Addressing the quality management by smallholder ware potato producers assures direct yield and income effects at the level of the target audience. It is fairly scale and income neutral, and is thus suitable for both larger and smaller non-mechanized farms that dominate the Sub-Saharan potato production outside of South Africa. Furthermore, it is within the direct influence sphere of ware potato producers themselves, making it applicable at individual level, without requiring any collective action or wider system change.

An added advantage of addressing seed potato quality issues at the level of ware potato producers is that it contributes directly to improving the awareness with regard to the importance of high seed potato quality. This is not only of essence for changing the own seed potato management practices of ware potato farmers, but may simultaneously contribute to an increased interest in high-quality seed potatoes from specialized producers. Within Kenya this effect has been observed clearly, especially there where producers were demonstrated simultaneously the effect of high-quality seed potatoes from specialized growers next to using positive selection and current farmer practice.

In situations, where there is limited scope for building a commercial specialized seed potato chain, improving the quality of self-supply seed potatoes can still be an option. In the large parts of Sub Sahara Africa where agricultural product marketing is a major challenge and market integration of producers is poor, commercial specialized seed potato production has proven to be difficult to realize. In countries like Kenya, Uganda and Ethiopia, where potato is considered by policy makers an important commercial and food-security crop, there are small specialized seed systems. In other countries, however, there is no specialized seed system whatsoever, and farmers rely entirely on informal seed multiplication systems and self-supply. In such systems improving the quality at ware potato production level can assure direct results, while building a specialized seed potato chain from the ground takes longer to have an effect on seed potato quality.

Finally variety selection programmes in Sub-Sahara Africa run parallel to a dynamic of informal landraces coming and going and having their share in the market. As long as a landrace is not formally recognized and named, it will not be multiplied by a formal seed potato system, as no basic seed would be available and few specialized seed multipliers would incorporate it in their enterprise. For such varieties quality management by ware potato producers is the only option to improve the quality of seed potatoes used.

7.3.4 Improving the specialized seed potato chain

When looking more closely at the specialized seed potato chain the poor availability of high quality seed potatoes becomes immediately apparent. The amounts produced in the three countries were at the time of the study inconsequential compared to the total theoretical demand for seed potatoes when considering the areas planted with the crop. In all three countries, the production of starter seed (breeder, foundation, pre-basic and basic seed) is assured by public research organizations. These public organizations do not have the capacity nor the mandate to enlarge their production to a commercial scale, let alone organize the distribution system required to assure that well-sprouted seed potatoes of the right variety are at the right place at the right time against an affordable price.

Commercial investment and know-how in commercial seed multiplication and marketing would be very welcome to increase the availability of high-quality seed potatoes. With commercial investment in the production and marketing of seed potatoes, the current low proportion of seed potatoes deriving from the specialized seed system can be increased. Especially producers with a fairly reliable demand for their produce are in a good position to invest in intensification of their production. This does require efforts to improve awareness of producers about the economic benefit of this investment. Furthermore efforts to provide producers with market security and access to finance can be of value. Finally realistic economic models providing producers with advice on the best frequency to renew their seed stock, depending on expected yields and prices, would be an asset.

Technological advancements in seed potato multiplication may contribute to system improvement. It has to be recognized however, that there are both robust older technologies, based on clonal selection, as well as well-established rapid multiplication technologies available for seed potato multiplication. Appropriate technology as such is not the major bottleneck for specialized seed system advancement in Sub-Sahara Africa. The biggest challenge for the specialized seed potato chain is in the effective distribution of high-quality seed potatoes to those growers with the ability and desire to pay for them.

7.3.5 Improving the non-specialized seed potato production

The seed potato system analysis demonstrates that the current seed potato management practices of ware potato producers in Kenya, Uganda and Ethiopia are not contributing to quality maintenance across seed generations. Three different entry points for improvement can be identified: 1. seed borne disease management, 2. improved seed potato storage and preparation, and 3. varieties better adapted to farmer seed potato management practices.

Seed potato degeneration can be defined as the reduction of yield potential across successive generations of re-used seed. Seed borne diseases, of which viruses and bacterial wilt are considered most important, are the major cause of seed potato degeneration. Improved control of these diseases can reduce the degeneration rate of seed potato stocks. The use of insecticides could be recommended specifically for the control of persistent vector transmitted viruses such as PLRV (Struik and Wiersema, 1999). Such advice is tailored, however, to the needs of specialized seed producers, who obtain a premium price for high-quality seed potatoes with low virus infection rates. Ware potato producers only recycle a proportion of their crop as seed, but would have to spray their entire crop to control insect pests, which makes the economic benefit of sprays doubtful. Without further investigations into the costs and benefits of such insect vector control by ware potato farmers this cannot be recommended, for economic, environmental, and human health reasons.

A typical potato field in Eastern Africa is rather heterogeneous, as a result of soil fertility differences in the field, but also as a result of diseases. The current farmer practice is to select seed potatoes from the bulk of the harvest, without consideration for the health status. Positive selection, the identification of healthy-looking mother plants as the source of seed potatoes for the next season, was identified as a promising technology for quality maintenance by smallholder ware potato farmers, and was further investigated (Chapters 4, 5 and 6).

The small seed plot technique was identified as an option to address the issue of poor affordability of high quality seed for smallholders. Rather than replacing a large proportion of their seed stock, farmers can opt to invest in a small amount of high-quality seed potatoes, which they multiply once or twice themselves in a specifically designated nursery, free of soil borne diseases (Kinyua et al., 2001; Kinyua et al., 2005). Within this small nursery the specific extra care normally recommended for seed potato production (aphid control, watering, timely weeding, late blight control) is little time consuming. Especially in situations where bacterial wilt incidences are very high, routine replacement of the seed stock by smallholders with clean material sourced from disease free small seed plots could be preferable over positive selection from badly infected fields. Participatory farmer managed trials and researcher controlled trials to assess the merits of the small seed plot system, compared to positive selection and blind seed recycling, with different rotation crops under high bacterial wilt incidences were implemented. The results of the researcher controlled trials seemed to confirm that injecting clean seed suppressed bacterial wilt more than positive selection, which still had a measurable effect on incidences. Results are not reported here as an inundation in the third season of the trial spread the disease across the entire experimental field, thus preventing a final yield comparison between the differently treated plots.

The current seed potato storage practices by producers are largely sub-optimal. To assure seed potatoes with multiple strong sprouts they require diffuse light from the moment they emerge from dormancy (Rhoades and Booth, 1982; Struik and Wiersema, 1999). When stored in the dark, fewer, longer and weaker sprouts are produced, which has a direct adverse effect on their vigour at the moment of planting. In addition, producers, especially those in Kenya and Uganda, grow two or three crops of potato per year. To assure their seed tubers are sprouted at the moment of planting, they apply techniques for breaking the dormancy of the tubers. An optimization of these techniques could also contribute to better prepared seed tubers and thus more vigorous growth during the first weeks after planting. A simple measure such as the replacement of one corrugated iron sheet with a transparent sheet in the roof of mixed maize-potato stores of Kenyan farmers (Gildemacher et al., 2007a), or purposely constructed stores made of local materials in Uganda and Ethiopia (Gebremedhin et al., 2006) can already make an important difference.

A last entry point for improving the quality of seed potatoes produced by ware potato farmers would be to adapt varieties to the practices of producers, rather than the other way around. Virus resistance and tolerance is in this regard an important trait (Radcliffe and Ragsdale, 2002). In pre-breeding extreme resistance to PVY and PVX, and tolerance to PLRV can be crossed into the genome (Salazar, 1996). Genetic engineering technologies targeted at the transfer of the genes responsible for this resistance has effectively been used to create lines with the combined resistant to the three viruses (Radcliffe and Ragsdale, 2002). A very short dormancy is not looked at favourably by potato breeders as it reduces the shelf life, but it is highly appreciated by smallholders in East Africa (Wachira et al., 2008) as it facilitates greatly the use of seed potatoes from the own harvest. Other useful variety characteristics, not looked at in variety selection processes, would be the development of multiple strong sprouts in spite of poor storage conditions. A last characteristic appreciated by farmers in East Africa is the production of both larger and smaller tubers on the same plant, so that there are tubers suited as seed for the next crop, and large tubers for marketing.

The most important disadvantage of targeting ware potato producers with measures to improve the seed potato system is that the group that needs to change its cultivation practices is much larger than that of seed potato producers in the case of a seed system modelled after Western potato growing countries. This study indicates that there are gaps in the knowledge of ware potato farmers and agricultural advisors alike. Testing and adapting technologies that can assist producers in improving their seed potato quality maintenance may not be the biggest challenge. To assure impact at scale, investments are needed to improve the awareness and knowledge of a sizeable proportion of potato producers.

7.4 Positive seed potato selection

In response to the findings of the seed potato system analysis further emphasis was given to positive selection as a potentially suitable technology to contribute to seed potato quality management by smallholder ware potato producers in East Africa.

The first work done on positive selection, presented in Chapter 4, demonstrated convincingly that positive selection provides important chances for smallholders to improve their income from potatoes without having to invest additional cash. Chapter 5 shows that positive selection results in substantial reductions in PVY, PLRV and PVX infection levels, while a similar effect on the infection rates by other viruses prevalent in Kenya, such as PVS, PVM and PVA (Muthomi et al., 2009) is most likely. Chapter 4 showed in addition that the technology is of use in keeping bacterial wilt infections in potato plant populations under control. Chapter 6 shows that it is possible to promote the use of this technology by smallholders in a cost effective manner and that initial adoption rates by producers introduced to the technology through training are promising.

7.4.1 What makes positive selection attractive for smallholder producers?

There are several characteristics that make the technology attractive to smallholder producers, and likely to be successful when introduced on a large scale. Maybe the most important characteristic of the technology is that it can be understood intuitively. It is logical that good looking potato plants are likely to produce healthy seed potato tubers. The only difficulty is that at harvest time senescence of the plants has made it impossible to assess the health status of plants. Furthermore the vast majority of producers and advisory service providers in Kenya alike were of the perception that heterogeneity between potato plants in a field was a normal feature, not realizing that this was caused by yield-reducing virus infections. Typically one potato farmer mentioned to a researcher, after the season-long training had ended: "When you first came, we thought you could not teach us anything about potatoes. We thought we knew all about growing potatoes, as we have always been growing potatoes, and our fathers before us. Only now we know that most of our potato plants were sick."

A second important element contributing to the potential of positive selection is that its effect can be demonstrated visually in the own environment of potato producers, within a single season, as shown in the Chapters 4 and 6. The methodology of learning by doing used by the ware potato producers, within a field of their own, with resources farmers own themselves, assists in demonstrating that the technology is within their reach and can fit their farming practice. A third beneficial characteristic of positive selection is that it is not difficult to apply. The accurate recognition of disease symptoms proved to be fairly complicated for potato producers (see Chapter 4). However, the selection of the best looking plants in a field of potato plants turned out to be much easier. In addition, also when the selection is not performed in the best possible manner, still results can be obtained. Altogether, the technology is easy to adopt. It works without any collective effort, no additional resources other than sticks and the own labour for marking plants are required and no adaptation of the current practice, other than harvesting marked plants before the bulk of potato plants is needed. Finally, it was demonstrated in Chapter 5 that the technology is fully complimentary to the intermittent use of high-quality seed. This makes it easy to integrate it into a system in which potato producers do at intervals renew (part of) their seed potato stock from a reliable source of high-quality seed.

These advantages make that positive seed potato selection can be recommended as a 'must include' element of any training for smallholder potato producers in countries where sourcing of seed potatoes from the own seed stock is common practice. Seed selection from the own harvest is practiced by smallholder potato producers the world over, and positive seed selection can contribute to improving seed potato quality management by these producers. An exception would be those areas where there is only a single cold season, followed by a long hot period in which potatoes cannot be grown, and seed potatoes cannot be stored under ambient temperatures. Such conditions are, for example, found in the Sahelian zone of Africa.

7.4.2 Limitations of positive selection

There are, after all the advantages of positive selection, some limitations that require to be highlighted. In the first place, positive selection is not the technology that maximizes yield. Continuous renewal of the seed stock with high-quality tubers is likely to result in lower disease levels in the plant population than positive selection, and can lead to yields closer to the field yield potential of the variety grown. For producers seeking profit maximization, rather than optimizing return on investment, such as producers able to invest and with reliable market access, positive selection may not be the optimal seed quality management option. This being said, the results in Chapter 5 suggest that even when renewing the seed stock with high-quality seed once in every two seasons, which in the case of Kenya is a rare, high frequency of replacement (Chapter 3), positive selection would be a recommended technology to select seed potatoes for the one cycle of re-using seed from the own harvest. Currently, however, seeking maximum profit by intensification is not a strategy that is suited for the majority of smallholder producers. Smallholders are cash short and, with good reason, risk averse, and rather aim for maximizing profit per unit cash invested. Positive selection is a suitable technology for this.

Another important limitation is that positive selection is not a technology suitable for seed multiplication. Considering the low multiplication rate of a potato plant of roughly a factor 10 (Struik and Wiersema, 1999), farmers can only use positive selection to renew their own seed stock. When they would want to consider growing high-quality seed potatoes to sell to others, other strategies need to be applied to bulk up larger volumes of high-quality seed.

This means that the effect of applying positive selection does not go beyond the individual adopter. As such it is essential to expose and train a large number of producers to the technology, for it to have a sizeable effect on the general quality of seed potatoes planted in a country. Chapter 6 shows that the training of potato producers through farmer managed demonstration trials turned out to be effective and affordable in Kenya. Currently the same training methodology is being applied in several African and some Latin American countries. As the technology is spreading well beyond the sphere of influence of the International Potato Center (CIP), monitoring the number of producers that have been exposed to training on positive selection is hardly possible, but to the best available knowledge 43,685 producers received training on the technology roughly following the method developed under this research trajectory (Chapter 6). Considering the large number of producers that would benefit greatly from mastering the technology, further optimization of the training methodology, and especially seeking further reduction in the cost per producer trained is important. Searching for effective farmer-to-farmer training approaches and the use of mass media and public demonstration of the technology to complement the farmer group training are opportunities for further development of the training methodology.

7.5 Consequences for the future of seed potato systems in Sub-Sahara Africa

The insights gained in the research trajectory presented in this thesis can provide direction to future development and research interventions in the field of to improve seed potato systems in sub-Sahara Africa.

7.5.1 Consequences for seed potato system development interventions

In decision making on seed system interventions more consideration can be given to the balance between investing in the building of specialized seed multiplication and investing in the training of ware potato producers in the maintenance of their own seed stock. The decision making and advice by agricultural advisors can become better tailored to the specific circumstances of different producers. Producers with limited risks as a result of market security and sure rainfall or irrigation and access to cash or credit are well placed to regularly invest in high-quality seed potatoes from a specialized producer to maximize their profits. It has to be realized that this solution only provides service to a part of the potato production producers. Investing simultaneously in the on-farm quality maintenance of smallholder producers can provide the much needed additional dynamic required to also increase yields and thus surpluses of producers in a less favourable market, rainfall and cash availability position.

7.5.2 Consequences for further seed potato systems research

'How can smallholder potato productivity and profitability be increased in Kenya, Uganda and Ethiopia?', was the main question asked at the beginning of this thesis. The research results presented in response to this question contribute to answering this question for the three countries, with effects well beyond the region of East Africa (Chapter 6). There are, however, remaining, and newly surfaced uncertainties and questions surrounding seed potato systems in general, and positive selection more specifically. Several research directions can be suggested to further the understanding of seed potato system dynamics and to investigate opportunities for seed potato system improvement.

Further research is suggested to test the maximum potential of positive selection in smallholder production systems. The research presented here only compared positive selection to farmer selection over a single generation. A highly relevant question is how the technology compares over a number of generations to regular renewal of quality seeds on the one hand, and common farmer practice of blind selection from the bulk of harvested potatoes on the other. Does the yield level go down over time to the same low level as without any selection? This seems unlikely, as it was observed that also Kenyan landraces, for which never clean seed has been injected in the system, do benefit from positive selection, hinting at the possibility of 'regeneration' of a seed stock as opposed to degeneration resulting from indiscriminate recycling of seed potatoes form the last crop.

To further investigate the hypothesis that both 'regeneration' and 'degeneration' is possible in a potato plant population a factorial trial under high natural virus pressure with the following treatments was suggested in Chapter 5:
a. Regular (farmer selection), positive selection, and full renewal per season;
b. Clean and highly contaminated starter seed;
c. PVY, PVX resistance and PLRV tolerant variety versus very susceptible variety.
While implementing such factorial trials, it is essential to invest ample resources in monitoring intensively both the development of virus incidences, as well as the development of virus titer in the plants and tubers of the different treatments.

In addition, more applied research into additional technologies that can contribute to quality management of seed potatoes by ware potato farmers can further widen the options available. Examples are the seed plot technique (Kinyua et al., 2001; Kinyua et al., 2005), improved seed potato storage (Gebremedhin et al., 2006) and breaking dormancy using simple measures available to smallholder producers (Shibairo et al., 2006).

Finally it is felt that advances can be made in seed potato system modelling to improve the understanding of the dynamics of seed borne diseases, and especially viruses and bacterial wilt. If realistic parameters can be identified for modelling the behaviour of viruses and bacterial wilt over generations, better informed decisions can be made with regard to the use of resistance and tolerance traits in potato varieties. It will also assist in assessing the possible impact of different disease control measures. Such modelling efforts would allow for better predicting of the virus and bacterial wilt population dynamics at plant, field, but importantly, also entire potato system level. Modelling the behaviour of the disease, and its effect on yields opens the way to economic modeling and the development of decision making aids to decide on the most economic replacement frequencies of farmer seed stocks by high quality seed potatoes. At system level it could assist in decision making with regard to the focus of sector development interventions aimed at seed system improvement.

7.6 Lessons on the role of agricultural research in accelerating innovation

On the basis of the experiences gained in the research process presented in this thesis, it is possible to investigate more closely the role of agricultural research in the deliberate process of agricultural innovation. Chapter 6 looked specifically into the role of research in the positive selection research trajectory. Here the role of research is considered within the entire intervention aimed at potato system innovation described in this thesis, from the diagnostic phase to scaling-up of findings from the positive selection research is considered.

7.6.1 Some innovation concepts applied to the potato system research

A brief definition of innovation useful here is that "Innovations are new ideas, practices, or products that are successfully introduced into economic or social processes" (Kwadwo et al., 2008). In the research described in this thesis, the aim was not innovation as such, but specifically innovation in agriculture with the objective of poverty alleviation.

Often different types of innovation are recognized, to bring some order in the wide diversity of changes that can be lumped under the heading of innovation. The improvements we call innovation can be of technical, managerial, institutional or policy nature (Worldbank, 2007a). Kwadwo et al. (2008) distinguished between technical, organizational and institutional innovation. It has to be kept in mind however, that there is in reality no strict line between different types of innovation. Innovation, whether dominantly technical, organizational or institutional, most often requires a successful combination of changes in hardware (technological practices), software (knowledge and mindset required to make it work), and orgware (forms of organization, rules of interaction, and norms) (Leeuwis and Aarts, 2011; Smits, 2002). The case of positive selection illustrates this clearly. Although positive selection is a typical technical entry point for innovation, knowledge change by researchers, agricultural advisors and producers was essential for success. Furthermore, to realize this knowledge change, new research-producer-advisory service interactions and partnerships were build. For the purpose of this research we only distinguish between dominantly technical innovation and dominantly institutional innovation, with the latter encompassing changes in the relationships between actors, as well as changes in the formal and informal 'rules of the game'.

In this thesis we specifically discussed innovation in the potato sector, an agricultural sub-sector. Figure 8.1 provides a schematic picture of the actors and their roles and interrelations in an agricultural sub-sector. Chain actors own the product at some stage of the production and marketing process and add value to it. Chain supporters, such as research, provide in one way or another direct services to the production to consumption process. Chain context actors influence the wider environment in which the sub-sector functions, such as infrastructure, policies and market regulations (KIT and IIRR, 2010). Dominantly technical innovation takes place at the level of chain actors in the form of the adoption of new practices, but most often requires changes at the level of chain supporters and the chain context to be effective. From the opportunities for potato system improvement identified in the system diagnosis presented in Chapter 2, integrated pest management (IPM) of late blight and bacterial wilt, soil fertility management and seed quality improvement can be considered dominantly technical entry points for innovation.

Organizational and institutional innovation concern a variety of changes in interactions between actors, and the formal and informal rules that guide this interaction. Farmers may organize themselves to engage in collective marketing, or a research organization may adopt new ways of enumerating its staff to improve its performance. In the case of this thesis, the suggested importance of a better representation of the interests potato producers would require institutional innovation. Similarly, a different appreciation within a research organisation for the contribution of researchers to development impact, compared to measuring performance based on scientific publications alone, could be considered institutional innovation.

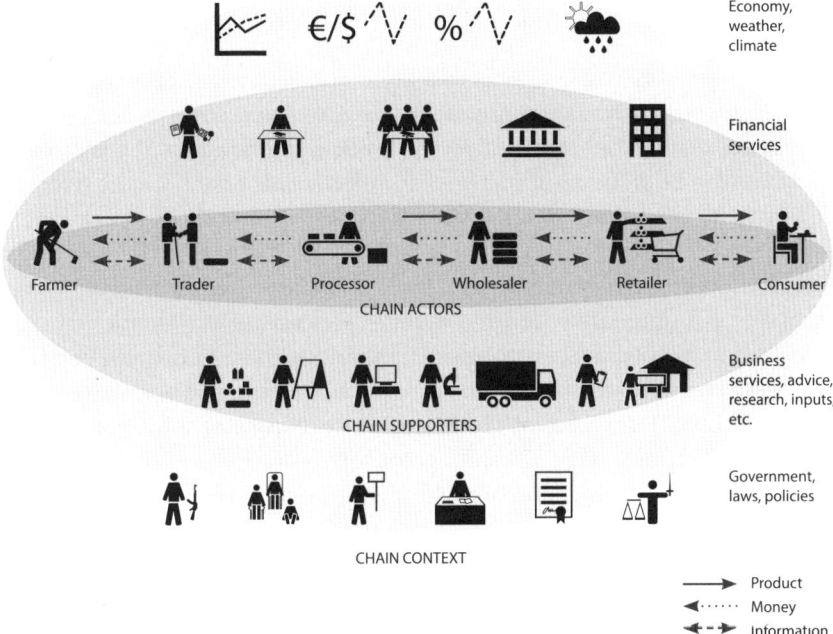

Figure 7.1: Schematic picture of an agricultural sub-sector (adapted from (KIT and IIRR, 2010)

7.6.2 The role of agricultural research in technical innovation

This thesis focused, after the first diagnostic chapter, on seed potato system improvement, and within the seed system on positive selection, which is predominantly a technical innovation. The research trajectory provides an opportunity to consider in retrospect the role of research, specifically in such innovations that have a dominantly technical entry point. To assess the role of research in technical innovation, the process of deliberate innovation can be divided into several elements, which will be discussed hereafter with a special focus on the role of agricultural research.

Inspired by the research trajectory described in this thesis a generic schematic presentation of the process of deliberate innovation aiming at poverty reduction was elaborated (Figure 7.2). The figure incorporates the insights obtained and lessons learned from the research trajectory. The deliberate agricultural innovation process contributing to the goal of poverty alleviation consists of four essential elements: opportunity assessment, the pilot innovation process, scaling-up and feedback into renewed efforts. As an optional element of the process there is researcher controlled testing, which can provide additional service to the process by generating opportunities from more fundamental research, or investigate specific issues rising from the pilot innovation process.

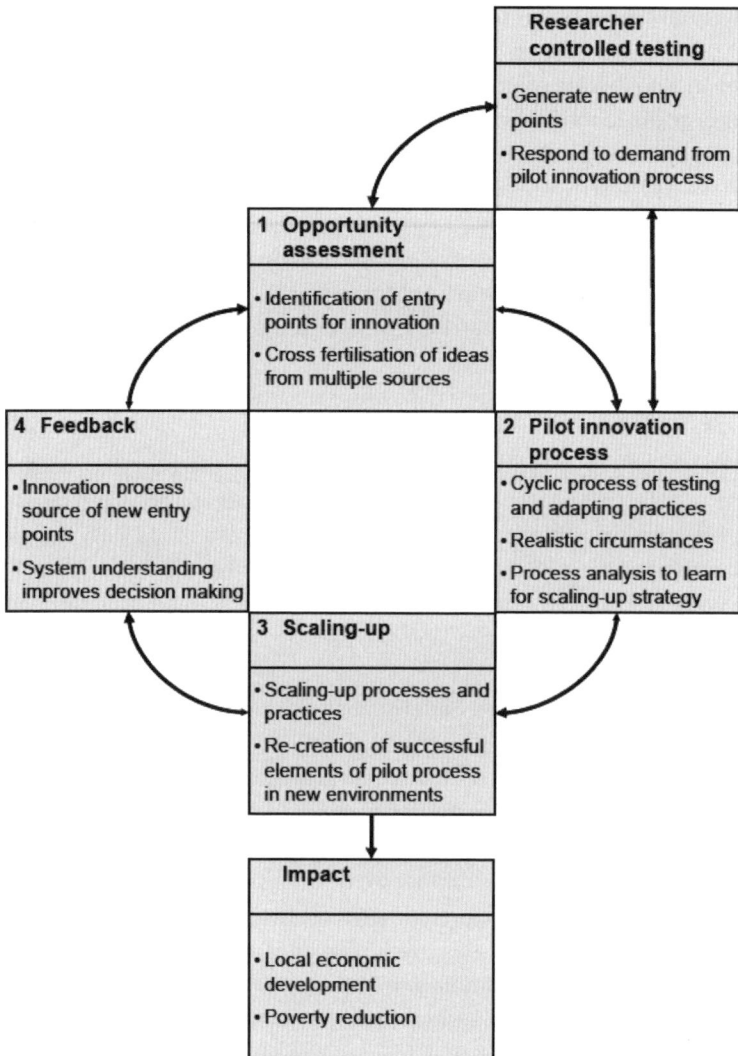

Figure 7.2: Schematic view of a deliberate innovation process aiming at poverty reduction.

Some caution is required in the use of the figure. The schematic picture is clearly an oversimplification of reality, as in reality the distinction between the elements in the process is not clear cut. It also has to be kept in mind that the different steps can be implemented simultaneously rather than subsequently, and a going back and forth between the steps is also possible (Van der Fliert and Braun, 2002). Furthermore, it is essential to remember that the figure shows the process of innovation, to which research can provide services. Research is not the unique contributor, nor necessarily the driver of the process (Hawkins et al., 2009). As stated by Hall et al. (Worldbank, 2007a): "Research is an important component – but not always the central component – of innovation". Finally the figure is by no means meant to refer back to the linear transfer of technology thinking, in which different entities are deemed responsible for steps in the process, and the transition from one step to the next is assumed to happen fairly spontaneously.

7.6.3 Opportunity assessment to identify entry points for innovation

The first step in a deliberate effort to trigger innovation is the assessment of problems and opportunities, or the identification of entry points for innovation. Van der Fliert and Braun (2002) also used this as the starting point for farmer participatory research. In this thesis, much emphasis has been given to this first step as described in Chapters 2 and 3. Research organizations dominated the needs and opportunity assessment by designing and implementing the quantitative surveys and facilitating the stakeholder workshops in the three countries. Although this work could be considered the capacity domain socio-economic scientists, technical researchers were equally involved.

The involvement of research in the needs and opportunity assessment and especially in the stakeholder workshops, provided both technical and socio-economic researchers with a unique opportunity to improve their understanding of the potato production system. Through the process they learned to appreciate the preoccupations and decision making rationale of the major economic actors and chain supporters. As such the result of the needs and opportunity assessment was two-fold: in the first place entry points for innovation were identified, but at the same time the actors involved improved their own understanding of the potato system, and the decision making rationale of its actors.

7.6.4 The pilot innovation process under realistic circumstances

On the basis of identified entry points for intervention, a pilot innovation process of testing and adapting opportunities for improvement can be initiated. Douthwaite (2002) described this as the 'learning selection' process, in which 'a plausible promise' is put into pilot practice and improved over a number of iterations, whereas Van der Fliert and Braun (2002) wrote about "innovation development" in their analysis of farmer participatory research processes. In the case of this thesis, the pilot innovation process of testing and adapting positive selection as a technology for seed quality maintenance by smallholders has been described in Chapters 4 and 6. Research played a leading role in this process. Research identified the 'plausible promise' that formed the entry point. Furthermore, research designed the process of testing and adapting, not only for the positive selection research discussed in Chapters 4 and 6, but also for parallel research on Integrated Pest Management (IPM) for the control of late blight and bacterial wilt which is not reported on this thesis.

The process of testing and adapting is by no means a monopoly mandate of agricultural research organisations. Testing and adapting options for improvement of farming practices is for many farmers and practitioners, such as agricultural advisors, part of their routine (Biggs, 1990; Chambers, 1983). Deliberate efforts to initiate technical innovation in agriculture, and specifically the process of testing and adapting practices may be led and run, without any research involvement, by farmers, possibly with the support of agricultural advisory services. It is felt, however, that even in such cases where research is not leading or driving the process, its advice on methodology of objective comparison between tested options, and its assistance in interpretation of results and joint reflection on required adaptation for a next iteration could be of added value.

The process of testing and adapting does not only serve to find the best technological fit in an experimental manner. At the same time it provides the opportunity to learn about the need for and also realize the changes in knowledge and organization that may be required for successful innovation. As Smits (2002) describes it, new knowledge (the invention in the words of Smith, or the plausible promise in the words of Douthwaite (2002)) should be seen as a potential opportunity that can be drawn upon, but more is required to transform it into successful applications. Only once knowledge is successfully applied, the invention has become innovation (Hall et al., 2005). As Leeuwis and Aarts (2011) indicate, innovation most often involves "the contextual re-ordering of relations in multiple social networks". Essentially this goes back to the same argument that for innovation to be successful, only change in the 'hardware' is often not enough and also changes in 'software' and 'orgware' are needed (Leeuwis and Aarts, 2011; Smits, 2002).

In the case of positive selection, the testing and adapting process was dominated by adapting the 'hardware' and 'software' components, while only limited changes in 'orgware' were found to be required. The technology in itself was of a simple nature in the sense that it could be adopted by individual famers without major re-organisation of social relationships, and it only needed to be verified to be working under high disease pressure in degenerated potato fields. The 'software' component meant that producers, but also advisory services, needed to realize that their current fields were degenerated by high virus incidence, and to learn they could actually improve that situation with their own hands. Also the mindset of research required change, as the notion that quality management by ware potato farmers could contribute to seed system improvement was not commonplace. The mindset of agricultural research organizations involved required adjustment through experiential learning and credible research evidence, much like the farmer and agricultural advisors. But even in the case of this simple technology, some institutional innovation in terms of building collaboration between international and national research and the Kenyan public extension, as well as mobilizing farmer self-help groups, was required to make positive selection a successful innovation.

To allow the process of testing and adapting to be effective in creating the fit between environment and innovative practice, the process has to take place under real enough, rather than in an artificial environment (Hommels et al., 2007). As was elaborated in Chapter 6, much effort was made in the case of positive selection to apply this principle to both the testing of the technology, as well as the testing and adapting of the training methodology. For assessing the technology of positive selection common producer circumstances were applied. Seed potatoes were sourced from an average, existing farmer field, of average quality and of a locally preferred variety. The demonstration trial was planted under farmer management practices. The actual selection of the mother plants in the source field was done by the producers themselves. Keeping the training circumstances realistic meant using the public extension staff as agents, training them in 2 days only, assuring allowances in compliance with their organizations' standards and assuring supervision through the public system, with only part-time supervision from public research.

7.6.5 Researcher controlled testing

The pilot innovation process is by nature a process that has to be implemented in a real enough life environment. This can, however, only be implemented once there is a 'plausible promise' or 'invention' that can be put to the test, or a limited number of technical options, to put to the test. Researcher controlled investigation, enquiry and testing, ranging from fundamental to more applied research, constitutes an important source, but not the unique source, of such plausible promises or inventions. Van der Fliert and Braun (2002), place fundamental and applied research within the process of farmer participatory research, and also in the 'farmer-back-to-farmer model' researcher controlled experimentation is placed explicitly within the innovation cycle. Here we prefer to place researcher controlled research separately, as an important optional, but not necessarily pivotal, service provider to the deliberate process of innovation, much like proposed by Hawkins et al. (2009) and in line with the idea of multiple sources of innovation (Biggs, 1990).

To come up with ideas that can be tested and adapted in a real life situation, often a narrowing down of options is required, which is easier and more effectively done in a systematic manner under researcher control. For example in the case of late blight IPM research, first a larger number of spray regime options were tested on-station for their potential with limited farmer participation (Nyongesa et al., 2005), to allow for narrowing down to the most promising treatments to be tested and adapted further under farmer management (Nyongesa et al., 2007).

In other cases, more fundamental research questions come up during the pilot innovation process, which would benefit from researcher controlled investigation. Such was the case with the positive selection research. The farmer controlled research showed the potential of the technology (Chapters 4 and 6), but further researcher controlled research was initiated to better understand the mechanisms behind it related to virus epidemiology (Chapter 5). Even more researcher controlled research can contribute to furthering the understanding of the mechanism behind positive selection, already applied and promoted successfully in practice.

7.6.6 Scaling-up for impact

The need for scaling-up beyond the pilot innovation process

The linear interpretation of roles of research and agricultural advisory services has been dismissed and was replaced by other theories (Leeuwis and Aarts, 2011; Sulaiman and Hall, 2002). These other theories put, in one way or the other, more focus on the dynamics of interaction between stakeholders and experiential learning from joint action as essential for innovation. The theories emphasize the importance of the parallel adaptation and development of technology, knowledge and organization. While acknowledging that this is important for innovation, and without wanting to revert back to linear (Sulaiman and Hall, 2002) thinking about roles and responsibilities in technology transfer, it has to be acknowledged that in the linear model, there was a clear focus, although simplistic, on the wider dissemination of new insights. Local success achieved by stakeholder interaction and joint learning is not enough to assure poverty impact at scale, beyond the direct participants in and beneficiaries of the pilot innovation process. In short, there is a need for scaling-up results from the pilot innovation process. Typologies of scaling-up and different terms as horizontal, vertical, functional, quantitative scaling up are used for different pathways towards the end goal of impact at scale (Menter et al., 2004). Here the term scaling-up is used as the general term for the combination of processes aimed at realizing poverty impact at scale.

The transfer of successful innovation

The simple transfer of a successful innovation from one context to another, through mere communication efforts by professional change agents, is hardly ever possible (Leeuwis and Aarts, 2011), as the innovation process is highly context specific (Hall, 2007). A local adaptation or co-evolution process to assure a fit between the environment and the proposed practices is almost always required. At the same time, however, this has to be done with an increased efficiency, wherever possible, compared with the pilot innovation process. The pilot innovation process is a deliberate discovery process. Much that is discovered in the pilot phase does not have to be re-discovered when scaling-up the results. When a similar process is initiated elsewhere, the lessons from the first discovery process can assist in making more precise and better adjusted choices at the start to assure a faster and more cost effective process.

Elements of the pilot innovation process may have been essential for local adaptation of the technology, assuring that actors have the proper knowledge to make it work and the institutional and organizational circumstances required for success. In other words, there will be elements of the pilot innovation process and local specificities that were essential for realizing the change in hardware, software and orgware that makes up the innovation. These elements of the pilot process would have to be identified and incorporated in the scaling-up strategy. The reflection on what worked and what did not work during the pilot innovation process is essential for effective scaling-up. In a similar fashion Lundy et al. (2005) emphasized the importance of continuous reflection, critical review and documentation of experiences, in view of future interventions. It can provide essential insights in how successful pilots can be scaled-up for larger than the pilot impact. This analysis could be considered a role for well-informed researchers, which have taken active part in the pilot innovation process. Sulaiman and Hall (2002) emphasized this capacity to learn from technical and institutional innovation processes as an desirable skill for a reinvented agricultural extension. As such, the pilot innovation process has two functions: in the first place testing and adapting opportunities for change, and as a second function testing and adapting of the process of change, focused on scaling-up.

The case of the positive selection training programme

If we consider the example of the positive selection training, the successes obtained depended on a number of conditions as described in Chapter 6. The ease of collaboration between research and extension, and the existence of a functioning infrastructure of public extension services surely contributed to the successful innovation. When implementing a similar programme in a different environment, for example without a public extension system with motivated staff at grassroots level, an adaptation of the training system would be required.

In the case of the positive selection programme, researchers were closely involved in the pilot innovation process. Where the pilot innovation process stopped and scaling-up started is, however, impossible to say. Both steps were in reality intertwined as described in Chapter 6. This was even a deliberate choice, as the question whether the technology positive selection could be disseminated cost-effectively was considered from the start just as essential as the technical potential in the hands of smallholder producers. This also shows that Figure 7.2 has to be read with caution. Different elements of the process can take place simultaneously and possibly also in reverse order, with feedback between the different elements, as also emphasized by Van der Fliert and Braun (2002) .

The actual experiences during the testing and adapting of both the technology of positive selection, as well as the training programme provided the basic knowledge that was used for the design and initiation of larger follow-up programs. The first experiences demonstrated convincingly that it was necessary to continue to implement demonstration trials with farmer groups as the basis for training. Not with the objective to collect more data and get better evidence of the functioning of the technology, nor to further adapt the technology, but as an effective method of experiential learning by potato producers. Furthermore it was observed that the training of extension staff in the field was essential as, much like the producers, they did hardly have any knowledge of potato virus diseases.

The question remains what the role of researchers is in this step of scaling-up. In the case of positive selection, researchers have been of key importance in the development and testing, adaptation and popularizing of the training methodology. There are those that would argue, following the linear paradigm of thinking about the mandate of research, extension and producers, rather than following a systems paradigm (Sulaiman and Hall, 2002), that scaling-up goes beyond the mandate of researchers and is the responsibility of advisory services. In the case of the positive selection programme this would have meant proving the merit of the technology, and leaving the development of an effective training methodology to agricultural advisory services.

The case of positive selection shows that taking responsibility as research beyond the collection of evidence on what works, and engaging in partnership with advisory services in the development of a scaling-up strategy can contribute to innovation. Clearly, research does have an added value in the development of scaling-up strategies to pursue maximum impact on poverty. Especially in the case of positive selection a strategy to train potato producers in a cost effective manner in maintaining the quality of their own seed stocks is essential, as it will only have an impact at scale as it is being practiced by a sizeable proportion of the producers. This involvement does not mean that research takes on the mandate of training large numbers of producers, but that research partners with advisory services and other stakeholders and contributes to the scaling-up process by co-designing, testing and adapting the approach.

7.6.7 Feedback

In Figure 8.2 the box of scaling up is connected through 'feedback' with the box of opportunity assessment. This signifies an important reason for research to go beyond researcher controlled experimentation, and engage in opportunity assessment, the pilot innovation process and scaling-up. Surely a first reason for research participation in such 'out-of-mandate' activities is its direct contribution to the quality of these processes. A second important reason, however, to advocate for research participation in 'out-of-mandate' activities is the feedback it provides from real life situations. By its participation in the three elements of the process, researchers learn from chain actors such as producers, traders, processors and retailers, but also from other chain supporters such as for example agricultural advisors, input dealers and financial service providers.

The participation in the innovation process can contribute to improved priority setting in research, a better understanding of the demands, opportunities and constraints faced by the average producer, and provide research with a better understanding of system constraints for the effective use of research derived insights. Bentley (1994) even suggest that this may be an over more important result of researcher-farmer collaboration than the direct result of farmer participatory research. Learning from active participation in the innovation process assists agricultural research to improve its service to the innovation process. It allows for a better adapted offer of opportunities in the first step of the process. In addition the ability of research to assist in interpretation of findings during the pilot innovation process improves. Finally the improved insight from active participation in the process assists researchers in the co-development, with other actors, of scaling-up strategies.

7.6.8 Additional considerations regarding the process of deliberate innovation

Figure 7.2 gives the suggestion of a logical and chronological order in events, which would be a mistake. An important part of the failure of the linear transfer of technology model to effectively support agricultural innovation was exactly this assumption, that there is a certain inevitable order in these steps, and that one component of the process can be ended satisfactory, and the next step can be implemented. And worse, it was assumed that different steps were the specific mandate of different types of organizations (Sulaiman and Hall, 2002), and that the transition of one step to the next would happen naturally and spontaneously. Figure 7.2 should be interpreted for what it is, a gross simplification of reality, made for the purpose of analyzing the role of research in a dominantly technical innovation process. The components of the process can gradually flow into the other or even occur more or less simultaneously as seen in the case of positive selection.

What can also be observed from the positive selection experience, is that the process of innovation can benefit from continuity in terms of the actors involved. The experiences gained in one element of the process feeds the decision making of the other elements. For example, the component of opportunity assessment does not only have as a result the explicit knowledge about entry points for change, which can be communicated to other actors involved in the pilot innovation process of testing and adapting practices and approaches. A just as important result of the opportunity assessment component is the knowledge and understanding of the system, and relationships and trust gained between different actors. This is largely tacit knowledge, and its transfer to other actors is as a consequence only partially possible. Similarly experiences from the testing and adaptation process are partly tacit knowledge, which is difficult to transfer.

Another factor to consider is process coordination and overview. Deliberate efforts to steer innovation in a certain direction, which is what is done in most agricultural development initiatives, does require a level of coordination. Without going into the details of what this coordination or facilitation would entail, it does add value to the process, related to the interaction and decision making between different contributors to joint action. In the case of the positive selection experience, this role was played by researchers. In other cases this role may be assumed by other actors. Researcher do however have a number of advantages to play this role in the case of the potato sector in East Africa, such as relative impartiality in the eyes of chain actors, system overview, multidisciplinary capacities within the organization and a relatively sure continued presence compared to for example NGO's and producer organisations (Gildemacher et al., 2007c).

7.6.9 Consequences for agricultural research

Based on the above discussion and the discussion in Chapter 6, some consequences for agricultural research can be derived to improve its contribution to agricultural innovation.

Creating space for generating and pursuing entry points for innovation

To optimize the contribution of research to innovation, it would be important to consider whether researchers have ample freedom to participate in identifying, testing and adapting entry points for innovation. Researcher liberty in decision making and flexibility in pursuing 'plausible promises' with partners fit poorly in the dominating funding mode of agricultural research which is project based, aiming for beforehand pre-defined results. In that regard the conclusion by Biggs (1989) that local researchers need to be provided with the flexibility necessary to engage in investigation processes with producers remains valid and relevant. It would be important for research organizations to assess how they can allow and even stimulate agricultural researchers identifying and pursuing entry-points for agricultural innovation with partners.

Appreciate both invention and innovation

In research the most important performance indicator is the documentation of new research insights in the specific form of scientific publications. This search for new insights is important for the process of agricultural innovation as it is a way to generate new entry points to pursue. However, it is not the only contribution of agricultural research to the process of innovation, as research also has a role to play in the three elements in the innovation process presented in Figure 7.2. This means that besides the direct results of research and development interventions, also the quality of the process of innovation should be considered in appreciating the performance.

Pilot innovation process under realistic circumstances

To assure a pilot innovation process leads to innovation, it has to be implemented in the most realistic circumstances possible. This does, however, have as a drawback that (researcher) control over the results is fairly limited. But without letting go of much of the control, and allowing for failure of practices being tried, the process of adaptation and change cannot take place. Providing room for failure is not necessarily obvious in project based research, nor in agricultural development projects. The need to show success to funders of agricultural research and development efforts can easily lead to overprotection and pampering of practices under investigation, leading to 'false positive conclusions' regarding the fit between environment and practice under investigation. At the end of the pilot, however, the practice under investigation has not been exposed to the 'realistic conditions' advocated for by Hommels et al. (2007), making scaling-up doubtful.

Capacities

The different roles of agricultural research in the process of innovation requires a broad combination of capacities. Most importantly inter-disciplinarity is needed at three different levels. In the first place research organization require to harbour expertise in widely different fields, from more fundamental technical research capacity to qualitative socio-economic investigation. Furthermore, as Rhoades and Booth (1982) already emphasized, harbouring multiple disciplines within a single organization alone is not enough, inter-disciplinarity is required with research teams working together on a specific intervention. In the case of this research thesis, a combination of applied technical expertise and social science expertise proved to be of use. Finally inter-disciplinarity is demanded from individual researchers contributing to the process of innovation. Technical scientists require knowledge of socio-economic reality and research methods, while also social scientists require technical background knowledge of the agricultural commodity they are working on.

Broadening the engagement of agricultural research

Agricultural research has much to contribute to the process of innovation, as demonstrated earlier. This goes beyond what is considered the core mandate of research under the linear paradigm of agricultural research and extension (Sulaiman and Hall, 2002). To contribute effectively to innovation, it is essential that the engagement of agricultural research is broadened to wider activities in the process of innovation in partnership with other actors in agricultural sub-sectors, be it chain actors, chain supporters or chain context actors. A wider engagement improves the understanding of the system by researchers and improves the ability to understand and respond to needs for its services. Furthermore it may help in avoiding 'strong network failure' (Woolthuis et al., 2005) of research, a state in which research is 'locked-in' in well-established partnerships only confirming, not critically reconsidering the existing paradigms (Hall, 2007). In other words, engagement in these wider activities, and specifically the resulting interaction with actors with different views, opinions and concerns assists in avoiding the development of 'professional blindness' amongst researchers that hampers innovation. The fact that quality management by ware potato producers and the simple technology of positive selection have till the research described in this thesis not played an important role in seed system improvement interventions could well be the result of such professional blindness. The emergence of innovative ideas is more likely to occur there where different opinions and ways of thinking interact. As such a broader engagement of research in the process of innovation is needed to remain dynamic and open to new, or even to very old, ideas.

References

Adjei-Nsiah, S., C. Leeuwis, K.E. Giller, O. Sakyi-Dawson, J. Cobbina, T.W. Kuyper, M. Abekoe, and W. Van Der Werf. 2004. Land tenure and differential soil fertility management practices among native and migrant farmers in Wenchi, Ghana: Implications for interdisciplinary action research. NJAS - Wageningen Journal of Life Sciences 52:331-348.

Anandajayasekeram, P. 2011. The role of agricultural r&d within the agricultural innovation systems framework, pp. 1-33 Agricultural R&D—Investing in Africa's Future: Analyzing Trends, Challenges, and Opportunities, Vol. Conference working paper 6. ASTI, FARA, Accra, Ghana.

Anderson, J., and L. Van Crowder. 2000. The present and future of public sector extension in Africa: contracting out or contracting in? Public administration and development 20:373-384.

Arif, M., U. Azhar, M. Arshad, Y. Zafar, S. Mansoor, and S. Asad. 2011. Engineering broad-spectrum resistance against RNA viruses in potato. Transgenic Research:1-9.

Arnold, E., and M. Bell. 2001. Some new ideas about research for development, p. 279-316, In DANIDA, ed. In: Partnership at the Leading Edge: A Danish Vision for Knowledge, Research and Development Danish Ministry of Foreign Affairs, Copenhagen.

Balogun, O.S., T. Teraoka, and Y. Kunimi. 2005. Influence of the host cultivar on disease and viral accumulation dynamics in tomato under mixed infection with Potato virus X and Tomato mosaic virus. Phytopathologia Mediterranea 44:29-37.

Barker, H., and J. Woodford. 1987. Unusually mild symptoms of potato leafroll virus in the progeny of late-infected mother plants. Potato Research 30:345-348.

Bawden, F.C., B. Kassanis, and F.M. Roberts. 1948. Studies on the Importance and Control of Potato Virus X. Annals of Applied Biology 35:250-265.

Bentley, J.W. 1994. Facts, fantasies, and failures of farmer participatory research. Agriculture and Human Values 11:140-150.

Biggs, S. 2007. Building on the positive: An actor innovation systems approach to finding and promoting pro poor natural resources institutional and technical innovations. International Journal of Agricultural Resources, Governance and Ecology 6:144-164.

Biggs, S., and H. Matsaert. 1999. An actor-oriented approach for strengthening research and development capabilities in natural resource systems. Public Administration and Development 19:231-262.

Biggs, S.D. 1989. Resource-poor farmer participation in research: a synthesis of experiences from nine national agricultural research systems. OFCOR Comparative Study Paper - ISNAR:35.

Biggs, S.D. 1990. A multiple source of innovation model of agricultural-research and technology promotion. World Development 18:1481-1499.

Bokx, J.d., and J.v.d. Want. 1987. Viruses of potatoes and seed-potato production. 2 ed. Pudoc, Wageningen.

Brandolini, A., P.D.S. Caligari, and H.A. Mendoza. 1992. Combining resistance to potato leafroll virus (PLRV) with immunity to potato viruses X and Y (PVX and PVY). Euphytica 61:37-42.

Bryan, J. 1983. On-farm seed improvement by the potato seed plot technique, pp. 13. International Potato Center, Lima.

Canadian_Food_Inspection_Agency. 2010. Requirements for the production of Pre-Elite seed potatoes from sources other than Nuclear Stock In C. F. I. Agency, (ed.) Directive D-97-11, 2nd revised ed. Canadian Food Inspection Agency -, Ottawa, Ontario.

Casper, R., and S. Meyer. 1981. Die Anwendung des ELISA-Verfahrens zum Nachweis pflanzenpathogener Viren. Nachrichtenblatt des Deutschen Pflanzenschutzdienstes: 33:49-54.

Chambers, R. 1983. Rural development : putting the last first Longman, London.

CIP-UPWARD. 2003. Farmer Field Schools: From IPM to Platforms for learning and Empowerment. CIP-UPWARD, ARD, Rockefeller Foundation, Los Baños.

CIP. 1996. Bacterial Wilt Training Manual Centro International de la Papa, Lima.

Crehan, K., and A. Von Oppen. 1988. Understandings of "development': an arena of struggle. The story of a development project in Zambia. Sociologia Ruralis 28:113-145.

Crissman, C.C., L.M. Crissman, and C. Carli. 1993. Seed potato systems in Kenya: a case study CIP, Lima.

Cromme, N., A.B. Prakash, N. Lutaladio, and F.O. Ezeta, (eds.) 2010. Strengthening potato value chains; Technical and policy options for developing countries, pp. 1-147. FAO, Rome, CFC, Amsterdam, Rome.

de Bokx, J.A. 1972. Spread of potato virus S. Potato Research 15:67-70.

Devaux, A., D. Horton, C. Velasco, G. Thiele, G. Lopez, T. Bernet, I. Reinoso, and M. Ordinola. 2009. Collective action for market chain innovation in the Andes. Food Policy 34:31-38.

Döring, T.F. 2011. Potential and Limitations of Plant Virus Epidemiology: Lessons from the Potato virus Y Pathosystem. Potato Research 54:341-354.

Dormon, E.N.A., C. Leeuwis, F.Y. Fiadjoe, O. Sakyi-Dawson, and A. van Huis. 2007. Creating space for innovation: the case of cocoa production in the Suhum-Kraboa-Coalter District of Ghana. International Journal of Agricultural Sustainability 5:232-246.

Douthwaite, B. 2002. Enabling innovation : a practical guide to understanding and fostering technological change Zed Books, London.

Edquist, C., and B. Johnson. 1997. Institutions and Organizations in Systems of Innovation, p. 41-63, In C. Edquist, ed. Systems of innovation; technologies, institutions and organizations. Pinter, London; Washington.

Engel, P. 1995. Facilitatign Innovation: an actor oriented approach and participatory methodology to improve innovative social practice in agriculture, Wageningen Agricultural University, Wageningen.

Engel, P.G.H. 1997. The social organization of innovation : a focus on stakeholder interaction Royal Tropical Institute, Amsterdam.

Eshetu, M., O.E. Ibrahim, and B. Etenesh. 2005. Improving potato seed tuber quality and producers' livelihoods in Hararghe, Eastern Ethiopia. Journal of New Seeds 7:31-56.

Ewing, E. 1981. Heat stress and the tuberization stimulus. American journal of potato research 58:31-49.

FAO. 2011. The state of food insecurity in the world; How does international price volatility affect domestic economies and food security? Food and Agricultural Organization of the United Nations, Rome.

Funtowicz, S.O., and J.R. Ravetz. 1993. Science for the post-normal age. Futures 25:739-755.

Gebremedhin, W., A. Solomon, K. Bekele, and E. Gebre. 2006. Participatory Potato Technology Development and Dissemination in Central Highlands of Ethiopia, pp. 124-131, In T. Amede, et al., (eds.) Integrated Natural Resource Management in Practice: Enabling Communities to Improve Mountain Livelihoods and Landscapes.

Geels, F., and R. Raven. 2006. Non-linearity and expectations in niche-development trajectories: Ups and downs in Dutch biogas development (1973-2003). Technology Analysis and Strategic Management 18:375-392.

Getachew, T., and A. Mela. 2000. The role of SHDI in potato seed production in Ethiopia: Experience from Alemaya integrated rural development project, pp. 109-112 5th African Potato Association Conference, Vol. 5. African Potato Association, Kampala, Uganda.

Gildemacher, P., P. Demo, P. Kinyae, M. Wakahiu, M. Nyongesa, and T. Zschocke. 2007a. Select the best: positive selection to improve farm saved seed potatoes; Trainers manual. International Potato Center, Nairobi.

Gildemacher, P., E. Schulte-Geldermann, D. Borus, P. Demo, P. Kinyae, P. Mundia, and P. Struik. 2011. Seed Potato Quality Improvement through Positive Selection by Smallholder Farmers in Kenya. Potato Research 54:253-266.

Gildemacher, P., P. Demo, I. Barker, W. Kaguongo, G. Woldegiorgis, W. Wagoire, M. Wakahiu, C. Leeuwis, and P. Struik. 2009a. A Description of Seed Potato Systems in Kenya, Uganda and Ethiopia. American journal of potato research 86:373-382.

Gildemacher, P., C. Leeuwis, P. Demo, P. Kinyae, P. Mundia, M. Nyongesa, and P. Struik. 2012. Dissecting a successful research-led innovation process: The case of positive seed potato selection in Kenya. International Journal of Technology Management and Sustainable Development. SUBMITTED.

Gildemacher, P., W. Kaguongo, O. Ortiz, A. Tesfaye, G. Woldegiorgis, W. Wagoire, R. Kakuhenzire, P. Kinyae, M. Nyongesa, P. Struik, and C. Leeuwis. 2009. Improving Potato Production in Kenya, Uganda and Ethiopia: A System Diagnosis. Potato Research 52:173-205.

Gildemacher, P.R., J. Mwangi, P. Demo, and I. Barker. 2007b. Prevalence of potato viruses in Kenya and consequences for seed potato system research and development, pp. 84-92, In A. Khalf-Allah, (ed.) 7th triennial African Potato Association conference. African Potato Association, Alexandria, Egypt.

Gildemacher, P.R., J. Landeo, R. Kakuhenzire, W. Wagoire, M. Nyongesa, M. Tessera, N. Bouwe, A. Bararyenya, B. Hakazimana, N. Senkesha, E. Gashabuka, J. Muhinyuza, G. Forbes, and B. Lemaga. 2007c. How to integrate resistant variety selection and spray regime research for IPM of potato late blight in Eastern and Central Africa, pp. 84-92 7th triennial African Potato Association conference. African Potato Association, Alexandria, Egypt.

Gildemacher, P.R., M. Tessera, W. Wagoire, R. Kakuhenzire, M. Nyongesa, N. Bouwe, A. Bararyenya, B. Hakazimana, N. Senkesha, E. Gashabuka, J. Muhinyuza, G. Forbes, O. Ortiz, and B. Lemaga. 2007d. The researcher field school on potato late blight: building regional research capacity in Eastern and Central Africa, pp. 313-320 7th triennial African Potato Association conference. African Potato Association, Alexandria, Egypt.

Gildemacher, P.R., P. Maina, M. Nyongesa, P. Kinyae, W. Gebremedhin, Y. Lema, B. Damene, T. Shiferaw, R. Kakuhenzire, I. Kashaija, C. Musoke, J. Mudiope, I. Kahiu, and O. Ortiz. 2009c. Participatory Analysis of the Potato Knowledge and Information System in Ethiopia, Kenya and Uganda, p. 153-167, In P. C. Sanginga, et al., eds. Innovation Africa: Enriching farmers' livelihoods. Earthscan, Sterling.

Godin, B. 2006. The linear model of innovation - The historical construction of an analytical framework. Science Technology & Human Values 31:639-667.

Hall, A. 2005. Capacity development for agricultural biotechnology in developing countries: An innovation systems view of what it is and how to develop it. Journal of International Development 17:611-630.

Hall, A. 2006. Publicprivate sector partnerships in an agricultural system of innovation: Concepts and challenges. International Journal of Technology Management & Sustainable Development 5:3-20.

Hall, A. 2007. The origins and implications of using innovation systems perspectives in the design and implementation of agricultural research projects: Some personal observations. UNU-MERIT, Maastricht.

Hall, A., L. Mytelka, and B. Oyeyinka. 2005. Innovation systems: implications for agricultural knowledge and practice. CGIAR.

Hall, A., G. Bockett, S. Taylor, M.V.K. Sivamohan, and N. Clark. 2001. Why research partnership really matter: Innovation theory, institutional arrangements and implications for developing new technology for the poor. World Development 29:783-797.

Haverkort, A. 1986. Forecasting national production improvement with the aid of a simulation model after the introduction of a seed potato production system in central Africa. Potato Research 29:119-130.

Hawkins, R., W. Heemskerk, R. Booth, J. Daane, A. Maatman, and A. Adekunle. 2009. Integrated Agricultural Research for Development (IAR4D). Forum for Agricultural Research in Africa (FARA) Sub-Saharan Africa Challenge Programme (SSA CP), Accra

Heemskerk, W., S. Nederlof, B. Wennink, and B. Shapland. 2008. Outsourcing agricultural advisory services : enhancing rural innovations in Sub-Saharan Africa Royal Tropical Institute (KIT), KIT Development, Policy and Practice, Amsterdam.

Hirpa, A., M. Meuwissen, A. Tesfaye, W. Lommen, A. Oude Lansink, A. Tsegaye, and P. Struik. 2010. Analysis of Seed Potato Systems in Ethiopia. American journal of potato research 87:537-552.

Hocdé, H., B. Triomphe, M. Faure, and M. Dulcire. 2008. From participation to partnership: A different way for researchers to accompany innovation porcesses - challenges and difficulties., p. 89-103, In P. C. Sanginga, et al., eds. Innovation Africa: Enriching farmers' livelyhoods. Earthscan, Sterling.

Hommels, A., P. Peters, and W.E. Bijker. 2007. Techno therapy or nurtured niches? Technology studies and the evaluation of radical innovations. Research Policy 36:1088-1099.

Howells, J. 2006. Intermediation and the role of intermediaries in innovation. Research Policy 35:715-728.

IFAD. 2010. Rural Poverty Report 2011: New realities, new challenges: new opportunities for tomorrow's generation. International Fund for Agricultural Development, Rome.

IMF. 2011. Sustainaing The Expansion, Washington.

Kakuhenzire, R., W. Wagoire, G. Kimoone, P. Gildemacher, B. Lemaga, and A. Ekwamu. 2007a. Adapting fungicide spray regime recommendations to farm level conditions by farmers in Uganda, pp. 92-101 7th Triennial African Potato Association conference, Vol. 7. APA, Alexandria, Egypt.

Kakuhenzire, R., G. Kimoone, W. Wagoire, P. Gildemacher, B. Lemaga, A. Ekwamu, and B. Mateeka. 2007b. Impact of farmers' selected IDM options on potato late blight control and yield, pp. 801-807 8th African Crop Science Conference, Vol. 8. African Crop Science Society, El-Minia, Egypt.

Karyeija, R.F., J.F. Kreuze, R.W. Gibson, and J.P.T. Valkonen. 2000. Synergistic interactions of a Potyvirus and a phloem-limited Crinivirus in sweet potato plants. Virology 269:26-36.

Kemp, R., A. Rip, and J. Schot. 2001. Constructing transition pathsthrough the management of niches, p. 269-299, In R. Garud and P. Karnoe, eds. Path dependence and creation. Lawrence Earlbaum, Mahwah.

Kinyua, Z.M., J.J. Smith, C. Lung'aho, M. Olanya, and S. Priou. 2001. On-farm successes and challenges of producing bacterial wilt-free tubers in seed plots in Kenya. African Crop Science Journal 9:279-285.

Kinyua, Z.M., M. Olanya, J.J. Smith, R. El-Bedewy, S.N. Kihara, R.K. Kakuhenzire, C. Crissman, and B. Lemaga. 2005. Seed-plot technique: empowerment of farmers in production of bacterial wilt-free seed potato in Kenya and Uganda Bacterial wilt disease and the Ralstonia solanacearum species complex.

KIT, and IIRR. 2010. Value chain finance: Beyond microfinance for rural entrepreneurs KIT publishers, Amsterdam; IIRR, Nairobi.

KIT, and CFC. 2011. From sorghum to shrimp: a journey through commodity projects KIT publishers, Amsterdam.

Klerkx, L., and C. Leeuwis. 2008a. Balancing multiple interests: Embedding innovation intermediation in the agricultural knowledge infrastructure. Technovation 28:364-378.

Klerkx, L., and C. Leeuwis. 2008b. Matching demand and supply in the agricultural knowledge infrastructure: Experiences with innovation intermediaries. Food Policy 33:260-276.

Klerkx, L., A. Hall, and C. Leeuwis. 2009. Strengthening agricultural innovation capacity: Are innovation brokers the answer? International Journal of Agricultural Resources, Governance and Ecology 8:409-438.

Kristjanson, P., R.S. Reid, N. Dickson, W.C. Clark, D. Romney, R. Puskur, S. MacMillan, and D. Grace. 2009. Linking international agricultural research knowledge with action for sustainable development. Proceedings of the National Academy of Sciences of the United States of America 106:5047-5052.

Kwadwo, A., K. Davis, and D. Aredo. 2008. Advancing Agriculture in Developing Countries through Knowledge and Innovation: Synopsis of an International Conference. International Food Policy Research Institute, Addis Abeba.

Kwambai, T.K., M.E. Omunyin, J.R. Okalebo, Z.M. Kinyua, and P. Gildemacher. 2011. Assessment of Potato Bacterial Wilt Disease Status in North Rift Valley of Kenya: A Survey, p. 449-456, In A. Bationo, et al., eds. Innovations as key to the green revolution in Africa: exploring the scientific facts. Springer Netherlands, Dordrecht.

Leeuwis, C. 2004. Fields of conflict and castles in the air. Some thoughts and observations on the role of communication in public sphere innovation processes. The Journal of Agricultural Education and Extension 10:63 - 76.

Leeuwis, C., and A. van den Ban. 2004. Communication for rural innovation : rethinking agricultural extension Blackwell Science, Oxford.

Leeuwis, C., and N. Aarts. 2011. Rethinking Communication in Innovation Processes: Creating Space for Change in Complex Systems. The Journal of Agricultural Education and Extension 17:21-36.

Long, N., and J.D. Van Der Ploeg. 1989. Demythologizing planned intervention: an actor perspective. Sociologia Ruralis 29:226-249.

Loorbach, D. 2007. Transition Management: New mode of governance for sustainable development, Erasmus University, Rotterdam.

Lundvall, B.A. 1992. National systems of innovation: towards a theory of innovation and interactive learning Pinter, London.

Lundy, M., M. Gottret, and J. Ashby. 2005. Learning alliances: An approach for building multi-stakeholder innovation systems. CGIAR.

Lutaladio, N., O. Ortiz, A. Haverkort, and D. Caldiz. 2009. Sustainable Potato Production; guidelines for Developing Countries FAO, Rome.

MacKinnon, J., and J. Munro. 1959. Comparative rates of movement of potato virus X into tubers and eyes of three potato varieties. American journal of potato research 36:410-413.

Malapi-Nelson, M., R.H. Wen, B.H. Ownley, and M.R. Hajimorad. 2009. Co-infection of soybean with Soybean mosaic virus and Alfalfa mosaic virus results in disease synergism and alteration in accumulation level of both viruses. Plant Disease 93:1259-1264.

Menter, H., S. Kaaria, N. Johnson, and J. Ashby. 2004. Scaling up, In D. Pachiko and S. Fajisaka, eds. Scaling up and out: achieving widespread impact through agricultural research. CIAT, Cali.

Miha, A., H. Rossel, and G. Atiri. 1993. Incidence and distribution of potato viruses in Plateau State, Nigeria. African crop Science Journal 1:131-138.

Munro, J. 1961. The importance of potato virus X. American journal of potato research 38:440-447.

Muthomi, J., J. Nyaga, F. Olubayu, J. Nderitu, J. Kabira, S. Kiretai, J. Auro, and M. Wakahiu. 2009. Incidence of aphid-transmitted viruses in farmer-based seed potato production in Kenya. Asian Journal of Plant Sciences 8:166-171.

Nganga, S., and F. Shideler. 1982. Potato Seed Production for Tropical Africa International Potato Center, Nairobi.

Nyongesa, M., P. Gildemacher, C. Lung'aho, and M. Wakahiu. 2007. Participatory evaluation of selected components of integrated control of potato late blight in Central Kenya, pp. 1975-1978 African Crop Science Conference, Vol. 8. African Crop Science Society, El-Minia, Egypt.

Nyongesa, M.W., P.R. Gildemacher, M.W. Wakahiu, P. Demo, and C. Lung'aho. 2005. Optimizing fungicide spraying regimes for integrated management of potato late blight in Kenya., pp. 337-342 African Crop Science Conference, Vol. 7. African Crop Science Society.

Pretty, J.N. 1994. Alternative Systems of Inquiry for a Sustainable Agriculture. IDS Bulletin 25.37-49.

Priou, S., O. Barea, H. Equise, and P. Aley. 2009. Participatory Training and Research for Integrated Management of Potato Bacterial Wilt: Adapted Guide for the Philippines. UPWARD, CIP, and NPRCRTC, Los Baños.

Radcliffe, E., and D. Ragsdale. 2002. Aphid-transmitted potato viruses: The importance of understanding vector biology. American journal of potato research 79:353-386.

Rajalahti, R., W. Janssen, and E. Pehu. 2008. Agricultural innovation systems: from diagnostics toward operational practice World Bank, Washington.

Reestman, A.J. 1946. De beteekenis van de virusziekten van de aardappel naar aanleiding van proeven met gekeurd en ongekeurd pootgoed. European Journal of Plant Pathology 52:97-118.

Reestman, A.J. 1970. Importance of the degree of virus infection for the production of ware potatoes. Potato Research 13:248-268.

Rhoades, R.E., and R.H. Booth. 1982. Farmer-back-to-farmer: A model for generating acceptable agricultural technology. Agricultural Administration 11:127-137.

Rogers, E. 1995, 2003. Diffusion of innovations Free Press, New York.

Salazar, L. 1996. Potato viruses and their control International Potato Center (CIP), Lima.

Satoh, K., T. Shimizu, H. Kondoh, A. Hiraguri, T. Sasaya, I.R. Choi, T. Omura, and S. Kikuchi. 2011. Relationship between symptoms and gene expression induced by the infection of three strains of Rice dwarf virus. PLoS ONE 6.

Schot, J., and F.W. Geels. 2008. Strategic niche management and sustainable innovation journeys: Theory, findings, research agenda, and policy. Technology Analysis and Strategic Management 20:537-554.

Schut, M., A. van Paassen, C. Leeuwis, S. Bos, W. Leonardo, and A. Lerner. 2011. Space for innovation for sustainable community-based biofuel production and use: Lessons learned for policy from Nhambita community, Mozambique. Energy Policy 39:5116-5128.

Scott, G.J., M.W. Rosegrant, and C. Ringler. 2000. Global projections for root and tuber crops to the year 2020. Food Policy 25:561-597.

Shibairo, S., P. Demo, J. Kabira, P. Gildemacher, E. Gachango, M. Menza, R. Nyankanga, G. Chemining'wa, and R. Narla. 2006. Effects of Gibberellic Acid (GA3) on Sprouting and Quality of Potato Seed Tubers in Diffused Light and Pit Storage Conditions. Journal of Biological Sciences 6:723-733.

Shiferaw, B., J. Okello, and V. Ratna Reddy. 2009. Challenges of Adoption and Adaptation of Land and Water Management: Options in Smallholder Agriculture: Synthesis of Lessons and Experiences, p. 285-276, In S. Wani, et al., eds. Rainfed agriculture : unlocking the potential. CAB International, Wallingford.

Smits, R. 2000. Innovatie in de universiteit Inaugurele rede Universiteit Utrecht. Universiteit Utrecht, Utrecht.

Smits, R. 2002. Innovation studies in the 21st century: Questions from a user's perspective. Technological Forecasting and Social Change 69:861-883.

Spielman, D.J., J. Ekboir, and K. Davis. 2009. The art and science of innovation systems inquiry: Applications to Sub-Saharan African agriculture. Technology in Society 31:399-405.

Struik, P.C., and S.G. Wiersema. 1999. Seed potato technology Wageningen Press, Wageningen.

Sulaiman, V.R., and A. Hall. 2002. Beyond technology dissemination: Reinventing agricultural extension. Outlook on Agriculture 31:225-233.

Tatineni, S., R.A. Graybosch, G.L. Hein, S.N. Wegulo, and R. French. Wheat cultivar-specific disease synergism and alteration of virus accumulation during co-infection with wheat streak mosaic virus and Triticum mosaic virus. Phytopathology 100:230-238.

Tindimubona, S., R. Kakuhenzire, J.J. Hakiza, W.W. Wagoire, and J. Beinamaryo. 2000. Informal production and dissemination of quality seed potato in Uganda. , pp. 99-104 5th African Potato Association Conference, Vol. 5. African Potato Association, Kampala, Uganda.

Torrance, L. 1992. Developments in methodology of plant virus detection. Netherlands Journal of Plant Pathology 98:21-28.

Turkensteen, L.J. 1987. Survey of diseases and pests in Africa: fungal and bacterial diseases. Acta Horticulturae:151-159.

UN-ESA. 2011. World Population Prospects: The 2010 Revision. [Online] http://esa.un.org/unpd/wpp/Analytical-Figures/htm/fig_11.htm (verified 21-02-2012).

UNCTAD. 2011. The least developed countries report 2011; The potential role of south-South cooperation for inclusive and sustainable development:194.

UNECA. 2010. Economic Report on Africa 2010: Promoting high-level sustainable growth to reduce unemployment in Africa. United Nations Economic Committee for Africa, Addis Abeba.

UNECA. 2011. Economic Report on Africa 2011; Governing development in Africa - the role of the state in economic transformation, Addis Abeba.

UNFPA. 2011. State of the world population 2011; People and posibilities in a world of 7 billion. United Nations Population Fund, New York.

Untiveros, M., S. Fuentes, and L.F. Salazar. 2007. Synergistic interaction of Sweet potato chlorotic stunt virus (Crinivirus) with carla-, cucumo-, ipomo-, and potyviruses infecting sweet potato. Plant Disease 91:669-676.

Van der Fliert, E. 1993. Integrated Pest Management: Farmer Field Schools Generate Sustainable Practices. A Case Study in Central Java Evaluating IPM Training., Agricultural University Wageningen, Wageningen.

Van der Fliert, E., and A.R. Braun. 2002. Conceptualizing integrative, farmer participatory research for sustainable agriculture: From opportunities to impact Agriculture and Human Values 19:25-38.

Wachira, K., P. Gildemacher, P. Demo, W. Wagoire, P. Kinyae, J. Andrade, K. Fuglie, and G. Thiele. 2008. Farmer practices and adoption of improved potato varieties in Kenya and Uganda, Lima.

Wakahiu, M.W., P.R. Gildemacher, Z.M. Kinyua, J.N. Kabira, A.W. Kimenju, and E.W. Mutitu. 2007. Occurrence of potato bacterial wilt caused by Ralstonia solanacearum in Kenya and opportunities for intervention., pp. 267-271 7th Triennial African Potato Association Conference. African Potato Association, Alexandria, Egypt.

Wani, S., T. Sreedevi, J. Rockström, and Y. Ramakrishna. 2009. Rainfed Agriculture – Past Trends and Future Prospects, p. 1-36, In S. Wani, et al., eds. Rainfed agriculture : unlocking the potential. CAB International, Wallingford.

Were, H.K., R.D. Narla, J.H. Nderitu, and H.L. Weidemann. 2003. The status of potato leafroll virus in Kenya. Journal of Plant Pathology 85:153-156.

Woolthuis, R.K., M. Lankhuizen, and V. Gilsing. 2005. A system failure framework for innovation policy design. Technovation 25:609-619.

Worldbank. 1996. The Worldbank participation sourcebook World Bank, Washington DC.

Worldbank. 2007a. Enhancing agricultural innovation; how to go beyond the strengthening of research systems World Bank, Washington.

Worldbank. 2007b. World Development Report 2008; Agriculture for Development The World Bank, Washington.

Zaag, D.v.d. 1987. Growing seed potatoes, p. 176-203, In J. d. Bokx, et al., eds. Viruses of potatoes and seed-potato production, 2nd edition. Pudoc, Wageningen.

Summary

Potato sector provides opportunities for poverty reduction in East Africa

Potato (*Solanum tuberosum* L.) is a crop with a high potential to contribute to poverty reduction in Eastern Africa through income increase and improved food security. First and foremost its productivity in terms of energy produced per hectare per day is the highest of all major arable crops, and almost double that of wheat and rice. During the food crisis potato prices, as the prices of other root and tuber crops, fluctuated much less than prices of major cereals. There is a steadily growing demand for potatoes in Eastern Africa, which is expected to continue. The prospect of a steady rise in potato consumption provides opportunities for production increase of potato without an erosion of farm gate prices, and in addition value addition opportunities in the snack food chain. Potatoes are largely grown by smallholder producers, and serve as a combined household food security and a cash crop.

Average current potato yields in Eastern Africa of 10.5 tonnes/ha are well below the world average, and also much below the yields that are observed in the fields of better performing smallholders. Over the last decades the growing demand for potatoes in Eastern Africa has been met by an increase in area under production. Productivity improvement is possible and needed as additional increase in area under potato cultivation will further shorten rotations which brings high risks of soil borne disease build-up. The general question can be asked how smallholder potato productivity and profitability can be increased in Kenya, Uganda and Ethiopia to respond to the increasing market demand.

Identification of entry point for potato sector innovation

First a diagnosis of the potato system of Kenya, Uganda and Ethiopia was made by combining quantitative surveys and qualitative stakeholder interaction meetings. Constraints in knowledge and information flow in the potato system and production related constraints were identified. This led to the identification of entry-points for innovation. Integrated management of bacterial wilt (*Ralstonia solanacearum*) and late blight (*Phytophthora infestans*), soil fertility management and improving seed potato quality were identified as technology based opportunities for innovation. Improvement of potato supply chains and improving the knowledge exchange in the sector were identified as more systemic opportunities for potato sector improvement.

Seed potato system analysis

A specific analysis of the seed potato system in Kenya, Uganda and Ethiopia was executed to identify innovative approaches to seed potato system improvement. The study confirms that both virus diseases and bacterial wilt are endemic and likely contributors to the low yields observed in the region. A study done in Kenya of the virus infection status of seed potato tubers sold in village markets only 3% was free of the major viruses PVY, PLRV, PVX and PVA. *Ralstonia solanacearum* was found in 74% of potato farms sampled in two Kenyan potato growing districts.

The study clearly demonstrates that specialised seed potato multipliers, who specifically produce to sell seed potatoes, only produce and market a fraction of seed potatoes used in Kenya, Uganda and Ethiopia. Less than 5% of the potato farmers in the three countries indicated to source seed potatoes from specialized multipliers. Farmer predominantly rely on seed potatoes from neighbours and their own farm-saved seed potatoes. Over 50% even rely entirely on farm-saved seed. The research results do reveal that this strategy by potato producers does make economic sense in the absence of affordable high quality seed potatoes combined with limited market security for consumption potatoes. Also in the future it is expected that the local seed potato chain based on neighbour and self-supply of seed potatoes will continue to play an important role in the potato systems in the three countries.

To improve the system the local and specialized chain need to be tackled simultaneously. To improve the local chain the seed potato quality management practices by ware potato farmers require to be improved. To improve the seed potato quality management capacities by smallholder ware potato producers training on seed quality maintenance and bacterial wilt and virus management is needed. Research is required to improve the understanding of the yield reducing effects of mixed virus infections under smallholder production circumstances. Furthermore research into virus resistance deserves specific attention as it has the potential to mitigate much of the yield reducing effects of the largely practiced indiscriminate seed potato recycling. Private investment in specialized multiplication could increase the volumes of high quality seed potatoes produced, and reduce prices. In absence of these investments research organizations should continue to engage in starter seed production and promote multiplication by small scale specialized multipliers.

Improving the quality self-supply seed potatoes through positive selection

In response to the identification of seed potato quality management by consumption potato farmers as a strategy complementary to specialised seed system building, a programme was initiated in Kenya to assess the value of positive selection for East African potato systems. Positive selection entails the selection of healthy looking potato plants, before senescence starts masking disease symptoms, as mother plants of seed potatoes for the next season. On-farm research was implemented to assess whether farmer-managed positive seed selection could improve potato yields. Positive selection gave an average yield increase in farmer managed trials of 34% after a single season of applying the technology. This corresponded under the prevailing farm gate potato prices in Kenya to an increase of 284 Euro in profit per hectare at an additional production cost of only 6 Euro. Positive selection can be an important alternative and complementary technology to regular seed replacement, especially in the context of imperfect rural economies characterized by high risks of production and insecure markets. It does not require cash investments and is thus accessible for all potato producers. It can also be applied where access to high quality seed is not guaranteed. The technology is also suitable for landraces and not recognized cultivars that cannot be multiplied formally. Finally the technology fits seamlessly within the seed systems of sub-Sahara Africa, which are dominated by self-supply and neighbour supply of seed potatoes.

Assessment of the mechanism behind positive selection

In response to the successful application of positive selection in farmer managed demonstration fields, questions were rising with regard to the mechanisms behind the effect of the technology. To investigate the supposition that the observed effect could largely be contributed to a reduction in virus infection rates replicated trials were initiated around Kenya under widely different circumstances, and with different sources of seed potatoes.

The effect of selecting seed potatoes from healthy-looking mother plants (positive selection) was compared with the common Kenyan farmer practice of selecting seed tubers from the harvested bulk of potatoes (farmer selection) was investigated in 18 replicated trials. The results demonstrated that positive selection assured lower incidence of viruses PLRV, PVY and PVX and higher progeny yields than farmer selection. Using seed tubers from positive selection clearly out-yielded using seeds from farmer selection irrespective of the agro-ecology, crop management, soil fertility and variety. Positive selection had an effect on yield irrespective of the quality of the starter seed used. The average yield increase resulting from the technology amounted to 30%. The virus incidence was 35% (PVY), 35% (PVX) and 39% (PLRV) lower for seed tubers from positive selection than for seed tubers from farmer selection. Regression analysis showed this reduction in virus incidence contributed to the higher yields as a result of positive selection, but did not fully account for the effect. It is likely that other, not tested, virus diseases such as PVM and PVA play a role, while also other seed borne diseases not analysed in this research may have played a role. On the basis of the research results it can be concluded that positive selection can benefit all smallholder potato producers who at some stage select seed potatoes from their own fields, and should thus be incorporated routinely in agricultural extension efforts.

Implications for seed potato system improvement

The combined research results lead to the conclusion that positive selection provides a promising complementary technology to improve seed potato systems in Sub Sahara Africa. In addition it has been proven that it is possible to train large numbers of smallholder potato producers cost effectively. This is essential as for a significant impact on the overall quality of seed potatoes used it is required that a large proportion of ware potato producers improves its seed potato management practices.

The technology of positive selection does need to be put in perspective. The regular replacement of the seed potato stock with high quality seed potatoes, multiplied by specialists form tested disease free starter material is likely to result in higher yields than practicing positive selection. On the other hand, the research results have shown that the technology of positive selection is also useful with a, for East African standards, very high replacement rate of once every two seasons.

The research results indicate that a careful consideration is needed in the design of seed potato system interventions for an effective balance between focussing on the specialised seed potato multiplication system and seed potato quality maintenance by ware potato producers. It is essential to consider the diverse needs of potato producers with different levels of market access and production scales.

Further research is suggested to test the maximum potential of positive selection in smallholder production systems, in combination with virus resistance. Modelling of seed potato quality at plant, field and system scale, combined with economic modelling, could contribute to improving decision making. Furthermore there are other technologies than positive selection that can contribute to seed potato quality management by smallholder ware potato producers. These technologies, such as seed potato nurseries, improved management of potato dormancy, sprouting and storage, can provide ware producers with additional options to maintain the quality of their seed potato stock.

Lessons from the research trajectory on the role of agricultural research in innovation

In retrospect the research trajectory can be considered as a successful contribution of agricultural research to innovation. As a result of the research effort an alternative parallel intervention strategy has opened up for the improvement of the potato sector in Sub Sahara Africa and the training approach developed under the program is now widely promoted in sub-Sahara Africa. The research on positive selection provides a case to investigate the role of agricultural research in agricultural innovation. The case is of particular interest because the entry-point was the re-use of an obsolete technology and the role of research went beyond the validation or falsification theoretical principles. The case points out that innovation can emerge from old technology, and is sometimes possible within the existing institutional environment. Placing innovation central, rather than research result, widens the role of research. In the case of positive selection research assumed responsibility, in partnership with extension, for the development and piloting of an effective training approach as primary objective, with research results being considered of secondary importance. Researchers could effectively contribute to innovation as a result of the provided and created freedom of to pursue a 'bright idea' and flexibility of manoeuvring. It is concluded that it is worthwhile to search for opportunities for incremental innovation that do not require complex institutional change. The positive selection experience shows that these opportunities can be of a surprising simple nature, and based on technology. Essential for researchers to contribute to innovation is room to manoeuvre and opportunity to immerse in practical collaborative partnerships with practitioners. Most importantly, the engagement and responsibility of research needs to be broadened and allow for the active engagement in training, communication and scaling-up.

Samenvatting

De aardappelsector geeft mogelijkheden voor armoedebestrijding in Oost-Afrika
Aardappel (*Solanum tuberosum* L.) is een gewas met een hoge potentie om bij te dragen aan armoedebestrijding in Oost-Afrika door verbeterde inkomens en voedselzekerheid, niet in de laatste plaats omdat de aardappel de hoogste energie- productie per hectare per dag genereert van alle belangrijke voedselgewassen, en bijvoorbeeld het dubbele van tarwe of rijst. Tijdens de voedselcrisis fluctueerden de prijzen van aardappels, net als die van andere wortel- en knolgewassen, veel minder dan de prijzen van granen. Verder is er een gestaag, en naar verwachting blijvend, groeiende vraag naar aardappels in Oost-Afrika. Een groeiende aardappelconsumptie creëert mogelijkheden voor productieverhoging zonder een sterke daling van de aardappelprijzen voor de boer. Daarnaast zijn er belangrijke kansen voor het vergroten van de toegevoegde waarde in de fast-food en de snack-food ketens. De teelt van aardappels wordt in Oost-Afrika gedomineerd door kleine boeren, en dient tegelijkertijd voor voedselvoorziening en het genereren van inkomen.

De huidige gemiddelde aardappelopbrengst in Oost-Afrika is slechts 10,5 ton per hectare en ligt beduidend lager dan het wereldwijde gemiddelde, en ligt ook fors onder de praktijkopbrengsten van de beter presterende kleine boeren. Gedurende de laatste decennia is vooral door het vergroten van het areaal aardappel voldaan aan de groeiende vraag. Nu is echter productiviteitsverhoging noodzakelijk om aan de groeiende vraag te voldoen, daar een verdere vergroting van het areaal aardappel de vruchtwisseling verder zou vernauwen hetgeen hoge risico's met zich meebrengt voor het opbouwen van bodemgebonden ziekten en plagen. De algemene vraag kan worden gesteld: hoe kan de productiviteit van aardappels van kleine boeren in Kenya, Uganda en Ethiopië duurzaam worden verhoogd om te kunnen voldoen aan de groeiende vraag?

Identificatie van mogelijkheden voor innovatie van de aardappelsector
Een eerste diagnose van het aardappelsysteem van Kenia, Oeganda en Ethiopië is gemaakt door een combinatie van kwantitatieve surveys en kwalitatieve, interactieve bijeenkomsten met belanghebbenden. Productie-gerelateerde problemen en belemmeringen in kennis en informatiestromen zijn in kaart gebracht. Op basis hiervan zijn mogelijkheden voor innovatie van de aardappelsector geïdentificeerd. Geïntegreerde bestrijding van bruinrot (*Ralstonia solanacearum*) en fytoftora (*Phytophthora infestans*), beheer van bodemvruchtbaarheid en verbetering van de pootgoedkwaliteit werden geïdentificeerd als

kansen voor technologische innovatie. De verbetering van aardappel-handelsketens en kennisuitwisseling in de sector werden geïdentificeerd als meer systemische kansen voor verbetering in de aardappelsector.

Analyse van het pootgoedsysteem

Om innovatieve manieren te zoeken om het pootgoedsysteem te verbeteren is een specifieke analyse uitgevoerd van het pootgoedsysteem in Kenia, Oeganda en Tanzania. De studie bevestigde dat zowel virusziekten als bruinrot endemisch zijn en waarschijnlijk bijdragen aan de lage aardappeloogsten die worden waargenomen in de regio. In een studie van de virusinfectie van op lokale Keniaanse markten als pootgoed verkochte aardappels bleek slechts 3% van de knollen vrij van de virussen PVY, PLRV, PVX en PVA. Bruinrot werd geconstateerd in 74% van de aardappelvelden in twee Keniaanse districten.

De studie toont overtuigend aan dat gespecialiseerde pootgoedvermeerderaars slechts een fractie van het in Kenia, Oeganda en Ethiopië gebruikte pootgoed produceren en vermarkten. Minder dan 5% van de aardappeltelers in de drie landen gaf aan pootgoed te betrekken van gespecialiseerde vermeerderaars. De meerderheid van de telers betrekt zijn pootgoed van buren en door selectie uit de eigen oogst. Meer dan 50% gebruikt zelfs uitsluitend zelf-geselecteerd pootgoed.

De onderzoeksresultaten hebben laten zien dat de door de kleine boeren toegepaste strategie van het gebruik van zelf-geselecteerde pootaardappel economisch gezien logisch is in een situatie waarin betaalbare, kwalitatief goede pootaardappelen ontbreken in combinatie met een beperkte marktzekerheid voor consumptieaardappelen. Ook voor de toekomst wordt verwacht dat de lokale keten op basis van burenhandel en zelfselectie van pootaardappelen een belangrijke rol blijft spelen in de aardappelsystemen van de drie landen.

Om het pootgoedsysteem te verbeteren moeten tegelijkertijd de lokale en de gespecialiseerde pootaardappelketen worden aangepakt. Om de lokale pootaardappelketen te verbeteren moeten consumptieaardappeltelers het beheer van pootgoedkwaliteit veranderen. Consumptieaardappeltelers hebben hiervoor behoefte aan training in het beheersen van virusziekten en bruinrot. Daarnaast is onderzoek noodzakelijk om de effecten van gemengde virusinfecties op de oogst onder omstandigheden van kleinschalige productie beter te begrijpen. Verder onderzoek naar virusresistentie verdient bijzondere aandacht omdat het de potentie heeft om de opbrengstverminderende effecten van het willekeurig hergebruiken van pootaardappelen te beperken. Commerciële investeringen in gespecialiseerde pootaardappelvermeerdering kunnen bijdragen aan het verhogen van het beschikbare volume van pootaardappelen van hoge kwaliteit en het verlagen van de pootgoedprijzen. Bij het uitblijven van deze investeringen moeten onderzoeksorganisaties basispootgoed blijven produceren en de verdere vermeerdering ervan door kleinschalige gespecialiseerde vermeerderaars blijven stimuleren.

Samenvatting

Het verbeteren van de kwaliteit van eigen pootgoed door positieve selectie

Verbeterd beheer van pootgoedkwaliteit door consumptieaardappeltelers werd geïdentificeerd als strategie complementair aan het verbeteren van het formele pootgoedvermeerderingssysteem. In dit licht werd een programma opgezet om de waarde van positieve selectie voor Oost-Afrikaanse consumptieaardappeltelers te onderzoeken. Positieve selectie bestaat uit het markeren van gezond ogende aardappelplanten, vóór het verouderingsproces inzet, om te dienen als moederplanten van pootgoed voor het volgend seizoen. In een on-farm onderzoeksprogramma werd bekeken of positieve selectie door consumptieaardappelproducenten hun oogsten zou kunnen verbeteren. In door boeren uitgevoerde proeven bleek een enkel seizoen positieve selectie de oogst met gemiddeld 34% te verhogen. Onder de toen geldende marketprijzen betekende dit een gemiddelde verbetering van het rendement met 284 Euro per hectare tegen een investering van slechts 6 Euro. Positieve selectie vormt dus een waardevolle technologie complementair aan het systeem van regelmatig gebruik van pootgoed van hoge kwaliteit van gespecialiseerde vermeerderaars, specifiek in de context van imperfecte rurale economieën gekarakteriseerd door hoge productierisico's en onzekere prijzen. Voor positieve selectie is geen geldelijke investering noodzakelijk en de technologie is dus toegankelijk voor alle producenten en kan ook daar worden toegepast waar toegang tot pootgoed van hoge kwaliteit niet is gegarandeerd. De technologie is ook bruikbaar voor landrassen en niet geregistreerde cultivars die niet formeel vermeerderd kunnen worden. Uiteindelijk past de technologie naadloos in het pootaardappelsysteem van Oost-Afrika, dat gedomineerd wordt door eigen vermeerdering en handel tussen buren.

Analyse van mechanismen achter positieve selectie

Als reactie op het succes van positieve selectie ontstonden vragen met betrekking tot de mechanismen die het effect zouden kunnen verklaren. Om de hypothese te toetsen dat de waargenomen effecten grotendeels konden worden toegeschreven aan een reductie in virusinfectieniveaus werden proeven met herhalingen opgezet door heel Kenia onder sterk verschillende omstandigheden, en met pootaardappels van verschillende bronnen.

Het effect van het selecteren van pootaardappels van gezond ogende moederplanten (positieve selectie) werd vergeleken met de gebruikelijke praktijk van het selecteren van pootgoed uit de bulk van geoogste aardappels (boeren selectie) in 18 proeven met herhalingen. De resultaten toonden aan dat positieve selectie heeft geleid tot een lagere infectie met de virussen PLRV, PVY en PVX dan de normale boerenselectie. Pootgoed verkregen via positieve selectie gaf ook een duidelijk hogere opbrengst dan pootgoed verkregen via boerenselectie, onafhankelijk van de agro-ecologie, teelttechniek, bodemvruchtbaarheid en cultivar. Positieve selectie had een opbrengstverhogend effect voor alle verschillende bronnen van pootgoed die werden gebruikt. De gemiddelde opbrengstverhoging was 30%. De virusinfectie was 35% (PVY), 35% (PVX) en 39% (PLRV) lager bij pootgoed verkregen via positieve selectie dan bij pootgoed verkregen via boeren-selectie. Regressieanalyse liet zien dat deze reductie in virusinfectie heeft bijgedragen aan de hogere opbrengst, maar dat deze niet het gehele oogsteffect leek te verklaren. Het is waarschijnlijk dat andere, hier niet geteste, virusziekten veroorzaakt door bijvoorbeeld PVM en PVA een rol hebben gespeeld, terwijl ook een effect van

173

positieve selectie op andere pootgoedgebonden ziekten of –plagen die niet werden geanalyseerd niet kan worden uitgesloten. Op basis van de onderzoeksresultaten kan worden geconcludeerd dat alle kleine aardappelproducenten die op enig moment pootgoed betrekken van hun eigen consumptieaardappelproductie kunnen profiteren van positieve selectie, en dat de technologie dus routinematig opgenomen dient te worden in landbouwvoorlichting aangaande aardappelteelt in Oost-Afrika.

Implicaties voor de verbetering van pootaardappelsystemen

De gecombineerde onderzoeksresultaten leiden tot de conclusie dat positieve selectie een veelbelovende en complementaire technologie is die kan bijdragen aan het verbeteren van pootaardappelsystemen in Sub-Sahara Afrika. Daarnaast is het bewezen dat het mogelijk is om grote aantallen kleine boeren op een kosten-efficiënte wijze te trainen in het gebruik van de technologie. Dit is essentieel omdat een groot aantal boeren de technologie moet gaan gebruiken voor het bereiken van een verbetering op grote schaal van de kwaliteit van gebruikt pootgoed.

De technologie dient wel in perspectief te worden geplaatst. De regelmatige vervanging van het pootgoed met pootaardappelen van hoge kwaliteit, vermeerderd door specialisten vanuit ziektevrij startmateriaal, geeft hogere opbrengsten dan puur te vertrouwen op positieve selectie. Aan de andere kant hebben de onderzoeksresultaten getoond dat positieve selectie ook zinvol is indien om het seizoen het pootgoed wordt vervangen, wat voor Oost-Afrikaanse standaard een zeer hoge frequentie is.

De onderzoeksresultaten impliceren dat in het ontwerp van interventies in pootaardappel-systemen een zorgvuldige afweging gemaakt dient te worden in de balans tussen investeren in het verbeteren van de gespecialiseerde pootgoed-vermeerdering en het handhaven van de pootgoedkwaliteit door consumptieaardappelproducenten. Het is daarbij van belang om de behoeften van aardappeltelers met verschillende markt-integratie en productieniveaus in beschouwing te nemen.

Verder onderzoek wordt voorgesteld om de maximale potentie van positieve selectie in productiesystemen van kleine boeren, in combinatie met virusresistentie, te toetsen. Voorts kan het modelleren van pootaardappelkwaliteit op plant-, veld- en systeemschaal, gecombineerd met economische modellen, bijdragen aan een verbetering van besluit-vorming met betrekking tot interventies. Andere technologieën, zoals eigen vermeerdering in zaaibedden en een verbeterd onder controle houden van kiemrust, spruiten en opslag van pootaardappels kunnen consumptieaardappelproducenten helpen in het behouden van een hoge kwaliteit pootgoed van hun eigen veld.

Lessen uit het onderzoekstraject met betrekking tot de rol van landbouwkundig onderzoek in innovatie

Terugkijkend kan het onderzoekstraject worden beschouwd als een succesvolle bijdrage van landbouwkundig onderzoek aan innovatie. Als resultaat van de onderzoeks-inspanning is er een nieuwe parallelle interventiestrategie beschikbaar gekomen voor de verbetering van pootaardappelsectoren in Sub-Sahara Afrika. De ontwikkelde trainings-methode wordt nu breed gebruikt en verder gepromoot. Het onderzoekstraject vormt een casus waarin de rol van landbouwkundig onderzoek in het proces van landbouwinnovatie kan worden onderzocht. De casus is van bijzondere waarde omdat de mogelijkheid voor innovatie bestond uit het hergebruik van een overbodig geachte technologie, en omdat de rol van onderzoek verder ging dan het valideren of falsificeren van hypotheses. Deze ervaring laat zien dat innovatie kan ontstaan op basis van oude technologie, en in sommige gevallen binnen de bestaande institutionele omgeving. Het centraal plaatsen van innovatie, in plaats van onderzoeksresultaten, verbreedt de rol van onderzoek. In het geval van dit onderzoekstraject heeft het onderzoek als belangrijkste doelstelling het ontwikkelen en populariseren van een effectieve trainingsaanpak genomen, met onderzoeksresultaten als secondaire doelstelling. Onderzoekers konden effectief aan het proces van innovatie bijdragen door de ruimte die zij kregen en creëerden om "slimme ideeën" na te jagen en de flexibiliteit in de uitvoering.

De conclusie kan zijn dat het de moeite loont om te zoeken naar mogelijkheden voor incrementele innovatie waarvoor geen complexe institutionele verandering noodzakelijk is. De ervaringen van positieve selectie laat zien dat deze mogelijkheden verrassend eenvoudig kunnen zijn. Essentieel voor onderzoekers om bij te dragen aan innovatie is beslissingsvrijheid en de mogelijkheid om samenwerking met uitvoerende partners aan te gaan. Meest belangrijk is dat het mandaat en de verantwoordelijkheid van onderzoek verbreedt wordt en er ruimte wordt gegeven voor actieve betrokkenheid bij training, communicatie en disseminatie.

Acknowledgements

First I would like to acknowledge the International Potato Center (CIP), for I which I worked from 2003 till 2007 with great pleasure. The CIP Nairobi office provided me with a hospitable and inspiring environment. Most importantly it provided to its scientific staff a large amount of freedom to deploy initiatives and to go there where there seemed to be opportunity and momentum for contributing to the improvement of potato farmer livelihoods. The collaboration with, and capacity building of local partners, impact on poverty and scientific results were equally valued, thanks to Charlie and later Jan.

The work presented here has two fathers. Paul, thanks a lot for the good collaboration and friendship and being with me at the origin of the work presented here, and continuing to be the best possible advocate of pragmatic solutions for resource poor potato farmers. Thanks also to Wachira for contributing to the work, and Dinah and Elmar for continuing where I left off. Thanks Lieven for being my friend and colleague in the office, Simont, Emily, Naomi, George, Eliud and Rueben for your support, and all other CIP Nairobi colleagues for making a great team.

CIP Lima and the other CIP regional offices provided an unimaginable reservoir of dedicated experts, indispensable for the multi-disciplinary challenges I faced in my work in East Africa. Oscar, Greg, Sylvie, Lucho, Graham, Juan, Merideth, Ian, Stef, Thomas, Carlos, André, Willy, Dindo, Mike, Fernando and others, thanks for sharing your insights and contributing to the success of the potato research and development work in East Africa.

I would like to specifically highlight the essential contributions by the numerous collaborators from national research stations, with whom I was working on a joint mission of improving potato systems in Eastern Africa. Thanks to Peter, Moses, Mercy, John, Jackson, Charles, Zacharia and all other contributors from KARI-NPRC and KARI-NARL. You were more colleagues than collaborators. Similar gratitude goes to the Uganda collaborators, William, Roger, George and Imelda in Kachwekano ZARDI, Ignatius of AFRICARE, Mike of CIP Kampala and Berga from PRAPACE. Geberemedhin, Mesfin, Agajie, Yohannes and colleagues from EIAR Holetta made me appreciate the dedication of an Ethiopian on a mission. Furthermore thanks to Ntizo and Eugène in Rwanda, Bernadette, Astère and Dieudonné in Burundi, Bouwe in DR Congo, Betty in Tanzania and Jean-Marc in Madagascar. You made my work possible, and taught me the essentials of potato systems in Eastern Africa.

I owe gratitude to all technicians, enumerators and students who have been involved in data collection. Special thanks to Pauline, Mwalimu, Thomas, Grace, John and Paul who I had the pleasure to work with during their MSc. thesis research work.

Thanks to the Ministry of Agriculture of Kenya, the coordinators at district and division offices and the frontline extension workers taking responsibility for delivering the training we co-designed and pilot tested. This is just as much the fruit of your work as it is of mine.

I have greatly appreciated the support I received from my colleagues at the Royal Tropical Institute that kept me motivated to continue working towards the final result, in spite of the other work that was on our hands. I also appreciate the WUR hosting me at the Haarweg for a period of writing on my thesis and will remember the welcoming attitude of the WUR staff and students at the Haarweg during my brief spell there.

The piece of work presented here would not have existed without the dedicated support by professors Paul Struik and Cees Leeuwis. Paul and Cees, thanks for the enriching discussions, your valuable contributions to the writing and your always constructive and motivating feedback. Thanks both for your endurance and supporting me to the end of the process.

Thanks to the thousands of potato farmers in East Africa, and in Kenya in particular who contributed to this work. You made it worth the effort. You made for my moments of euphoria, driving back home from the field after hard long days of work, knowing that together we were making a small, but significant difference.

Finally there was one person with me every step of the way. Anje, bedankt.

Publication list

Journal papers

Boekhout, T., P. Gildemacher, B. Theelen, W.H. Muller, B. Heijne, and M. Lutz. 2006. Extensive colonization of apples by smut anamorphs causes a new postharvest disorder. Fems Yeast Research 6:63-76.

Gildemacher P., F. van Alebeek and B. Heijne, 2001. Farming system comparison in integrated apple growing. Bulletin OILB/SROP 24(5), 21-26.

Gildemacher, P., B. Heijne, M. Silvestri, J. Houbraken, E. Hoekstra, B. Theelen, and T. Boekhout. 2006. Interactions between yeasts, fungicides and apple fruit russeting. Fems Yeast Research 6:1149-1156.

Gildemacher, P., C. Leeuwis, P. Demo, P. Kinyae, P. Mundia, M. Nyongesa, and P.C. Struik. 2012. Dissecting a successful research-led innovation process: The case of positive seed potato selection in Kenya. International Journal of Technology Management and Sustainable Development. SUBMITTED.

Gildemacher, P., E. Schulte-Geldermann, D. Borus, P. Demo, P. Kinyae, P. Mundia, and P.C. Struik. 2011. Seed Potato Quality Improvement through Positive Selection by Smallholder Farmers in Kenya. Potato Research 54.253-266.

Gildemacher, P., P. Demo, I. Barker, W. Kaguongo, G. Woldegiorgis, W. Wagoire, M. Wakahiu, C. Leeuwis, and P.C. Struik. 2009. A Description of Seed Potato Systems in Kenya, Uganda and Ethiopia. American Journal of Potato Research 86:373-382.

Gildemacher, P., W. Kaguongo, O. Ortiz, A. Tesfaye, G. Woldegiorgis, W. Wagoire, R. Kakuhenzire, P. Kinyae, M. Nyongesa, P. Struik, and C. Leeuwis. 2009. Improving Potato Production in Kenya, Uganda and Ethiopia: A System Diagnosis. Potato Research 52:173-205.

Ortiz, O., Orrego, R., Pradel, W., Gildemacher, P., Castillo, R., Otiniano, R., Gabriel, J., Vallejo, J., Torres, O., Woldegiorgis, G., Damene, B., Kakuhenzire, R., Kasahija, I. & Kahiu, I. Incentives and disincentives for stakeholder involvement in participatory research (PR): lessons from potato-related PR from Bolivia, Ethiopia, Peru and Uganda. International Journal of Agricultural Sustainability 9(4): 522-536.

Schulte-Geldermann, E., P.R. Gildemacher, and P.C. Struik. 2012. Improving seed health and seed performance by positive selection in three Kenyan potato varieties. American Journal of Potato Research (accepted with minor revision).

Shibairo, S., P. Demo, J. Kabira, P. Gildemacher, E. Gachango, M. Menza, R. Nyankanga, G. Chemining'wa, and R. Narla. 2006. Effects of Gibberellic Acid (GA3) on Sprouting and Quality of Potato Seed Tubers in Diffused Light and Pit Storage Conditions. Journal of Biological Sciences 6:723-733.

Conference papers and popular newsletters

Bouwe, N., P. Gildemacher, W. Wagoire, and B. Lemaga. 2007. Integrating fungicide applications and host resistance for control of potato late blight in Congo, pp. 92-101. 7th Triennial African Potato Association conference, Vol. 7. APA, Alexandria, Egypt.

Gildemacher, P., P. Demo, P. Kinyae, M. Nyongesa, and P. Mundia. 2007. Selecting the best plants to improve seed potato. LEISA Magazine 23:10-11.

Gildemacher, P.R., 2010. Potato system diagnosis in East Africa: An innovation system analysis, 85-91 In: Cromme, N., A.B. Prakash, N. Lutaladio, and F.O. Ezeta, (eds.). Strengthening potato value chains; Technical and policy options for developing countries. FAO, Rome and CFC, Amsterdam.

Gildemacher, P.R., J. Landeo, R. Kakuhenzire, W. Wagoire, M. Nyongesa, M. Tessera, N. Bouwe, A. Bararyenya, B. Hakazimana, N. Senkesha, E. Gashabuka, J. Muhinyuza, G. Forbes, and B. Lemaga. 2007. How to integrate resistant variety selection and spray regime research for IPM of potato late blight in Eastern and Central Africa, pp. 84-92. 7th Triennial African Potato Association conference. African Potato Association, Alexandria, Egypt.

Gildemacher, P.R., J. Mwangi, P. Demo, and I. Barker. 2007. Prevalence of potato viruses in Kenya and consequences for seed potato system research and development, pp. 238-241. 7th Triennial African Potato Association conference. African Potato Association, Alexandria, Egypt.

Gildemacher, P.R., K. Davis, N. Sellemna, W. Ochola and W. Heemskerk, 2011. Essential capacities for the future agricultural advisor and resulting challenges for education. Paper presented at the GFRAS International conference on innovations in extension and advisory services, 15-18 November 2011, Nairobi.

Gildemacher, P.R., M. Tessera, W. Wagoire, R. Kakuhenzire, M. Nyongesa, N. Bouwe, A. Bararyenya, B. Hakazimana, N. Senkesha, E. Gashabuka, J. Muhinyuza, G. Forbes, O. Ortiz, and B. Lemaga. 2007. The researcher field school on potato late blight: building regional research capacity in Eastern and Central Africa, pp. 313-320. 7th Triennial African Potato Association conference. African Potato Association, Alexandria, Egypt.

Gildemacher, P.R., B. Heijne, J. Houbraken, T. Vromans, S. Hoekstra, and T. Boekhout. 2004. Can phyllosphere yeasts explain the effect of scab fungicides on russeting of Elstar apples? European Journal of Plant Pathology 110:929-937.

Kaguongo W, Gildemacher P, Low J., 2007. Influence of market and farmer preferences on potato variety adoption, pp. 321–329. 7th triennial African Potato Association conference. African Potato Association, Alexandria, Egypt.

Kakuhenzire, R., G. Kimoone, W. Wagoire, P. Gildemacher, B. Lemaga, A. Ekwamu, and B. Mateeka. 2007. Impact of farmers' selected IDM options on potato late blight control and yield, pp. 801-807 8th African Crop Science Conference, Vol. 8. African Crop Science Society, El-Minia, Egypt.

Kakuhenzire, R., W. Wagoire, G. Kimoone, P. Gildemacher, B. Lemaga, and A. Ekwamu. 2007. Adapting fungicide spray regime recommendations to farm level conditions by farmers in Uganda, pp. 92-101. 7th Triennial African Potato Association conference, Vol. 7. APA, Alexandria, Egypt.

Nyongesa, M., P. Gildemacher, C. Lung'aho, and M. Wakahiu. 2007. Participatory evaluation of selected components of integrated control of potato late blight in Central Kenya, pp. 1975-1978. African Crop Science Conference, Vol. 8. African Crop Science Society, El-Minia, Egypt.

Nyongesa, M.W., P.R. Gildemacher, M.W. Wakahiu, P. Demo, and C. Lung'aho. 2005. Optimizing fungicide spraying regimes for integrated management of potato late blight in Kenya., pp. 337-342 African Crop Science Conference, Vol. 7. African Crop Science Society.

Ortiz, O., R. Orrego, R., W. Pradel, P. Gildemacher, R. Castillo, R. Otiniano, J. Gabrieli, J. Vallejo, O. Torres, G. Woldegiorgis, B. Damene, R. Kakuhenzire, I. Kashaija, I. Kahiu, 2007. Participatory Research on Potato-Related Innovation Systems in Bolivia, Ethiopia, Peru and Uganda. Paper presented at Farmer First Revisited Workshop, 12-14 December 2007.

Wakahiu, M.W., P.R. Gildemacher, Z.M. Kinyua, J.N. Kabira, A.W. Kimenju, and E.W. Mutitu. 2007. Occurrence of potato bacterial wilt caused by Ralstonia solanacearum in Kenya and opportunities for intervention., pp. 267-271. 7th Triennial African Potato Association Conference. African Potato Association, Alexandria, Egypt.

Books and book chapters

Gildemacher, P.R., Demo, P., Kinyae, P., Wakahiu, M., Nyongesa, M., Zschocke, T., 2007. Select the best; positive selection to improve farm saved seed potatoes. Trainers manual. CIP / CTA, / Wageningen, 108 pp. Reprinted, 2009. (*Also printed in French, Portuguese, Spanish, Swahili and Amharic*).

Gildemacher, P.R., Maina, P., Nyongesa, M., Kinyae, P., Gebremedhin, W., Lema, Y., Damene, B., Shiferaw, T., Kakuhenzire, R.,

Kashaija, I., Musoke, C., Mudiope, J., Kahiu, , Ortiz, O., 2006. Participatory analysis of the potato knowledge and information system in , and Uganda, pp. 153-166. In: Sanginga, P, Waters-Bayer, A., Kaaria, S., Njuki, J., Wettasinha, C. (eds.) Innovation : Enriching farmers' livelihoods.

Hawkins R., W. Heemskerk, R. Booth, J. Daane and A. Maatman, 2009. Integrated Agricultural Research for Development - IAR4D. A discussion draft for the Forum for Agricultural Research in Africa (FARA) Sub-Saharan Africa Challenge Programme (SSA-CP). With additional contributions from: Suzanne Nederlof, Toon Defoer, Nour Sellamna, Peter Gildemacher and Driek Enserink. FARA, ICRA, KIT.

KIT & CFC, 2011. From Sorghum to Shrimp: a journey through commodity projects. KIT publishers, Amsterdam. 150 pp.

Gildemacher, P.R., L. Oruku and E. Kamau-Mbuthia, 2012. Impact and sustainability, pp. 55-67 In: Nederlof, Wongtschowski and Van der Lee (eds.), Putting heads together; agricultural innovation platforms in practice. Bulletin 396, KIT publishers, Amsterdam.

Klerkx, L. and P.R. Gildemacher, 2012. The role of innovation brokers in agricultural innovation systems, pp 221-230 In: World Bank, Agricultural innovations: an investment sourcebook. World Bank, Washington.

Kwambai, T.K., M.E. Omunyin, J.R. Okalebo, Z.M. Kinyua, and P. Gildemacher, 2011. Assessment of Potato Bacterial Wilt Disease Status in North Rift Valley of Kenya: A Survey, pp. 449-456, In: A. Bationo, et al., (eds.) Innovations as key to the green revolution in Africa: exploring the scientific facts. Springer Netherlands, Dordrecht.

Wachira, K, P. Gildemacher, P. Demo, W. Wagoire, P. Kinyae, J. Andrade, K. Fuglie and G. Thiele, 2008. Farmer practices and adoption of improved potato varieties in Kenya and Uganda. In Farmer practices and adoption of improved potato varieties in Kenya and Uganda, Social Sciences Working paper, 85. CIP, Lima.

About the author

Peter Gildemacher, born in Assen, The Netherlands, 1975, obtained his BSc. and MSc. in Tropical Land Use with a specialization in Tropical Agronomy, Wageningen Agricultural University, in 1998. In 1996 he was attached as undergraduate to the Antenne Sahélliene in Burkina Faso to conduct participatory action research on agroforestry practices. In 1997 he conducted MSc. thesis research titled "Ethiopian Mustard (*Brassica carinata*) as a leafy vegetable in Tanzania; farmers' practices and possible improvements" at the Asian Vegetable Research and Development Center in Tengeru, Tanzania. A second thesis research entitled "Top dressing organic onions with liquid manure" was conducted in 1998 at the Department of Organic Farming, Wageningen Agricultural University, in the Netherlands.

Mr. Gildemacher was based in Nairobi, Kenya at the International Potato Center (CIP) from 2003 till 2007 and developed, managed and luated regional research and development projects in Eastern Africa. He has developed and effectively promoted a comprehensive extension program for improving seed potato quality in Kenya, which is currently being scaled up to Uganda, Ethiopia, Rwanda, Malawi, Mozambique and Angola. Furthermore he built regional research and local stakeholder interaction in Kenya, Uganda, Ethiopia, Rwanda, Burundi and DR Congo. Based on this work in East Africa this PhD thesis was elaborated under the supervision of the Centre for Crop Systems Analysis and the Communication and Innovation Studies group of Wageningen University.

Before that he worked as a technical assistant at the Centre National de Semences Forestières (CNSF) in Burkina Faso from 2000-2003, facilitating multi-stakeholder projects at a regional station of the institute, linking research to extension organizations. At the start of his career, in 1999 and 2000, he was employed by the Fruit Research Station in the Netherlands, managing applied research projects in crop protection in collaboration with extension organizations and producers.

Since 2008 mr. Gildemacher is working as senior advisor sustainable economic development at KIT (the Royal Tropical Institute). He designs, evaluates and advises in the implementation of complex agricultural development programs. He leads action research processes, documents knowledge from development practice and facilitates reflection and capitalisation efforts.

PE&RC PhD Education Certificate

With the educational activities listed below the PhD candidate has complied with the educational requirements set by the C.T. de Wit Graduate School for Production Ecology and Resource Conservation (PE&RC) which comprises of a minimum total of 32 ECTS (= 22 weeks of activities)

Review of literature (5.6 ECTS)
- Review of mainly technical literature has been done at the start of the assignment in Kenya; the socio-economic literature has been reviewed as part of the proposal writing process.

Writing of project proposal (4.5 ECTS)
- The PhD proposal was written by the PhD candidate entirely

Post-graduate courses (2.9 ECTS)
- Field research methods: methods and tools for qualitative data analysis; Mansholt, Wageningen (2008)
- Introduction to R for statistical analysis; PE&RC, Wageningen (2008)

Laboratory training and working visits (4.8 ECTS)
- Working visit CIP headquarters in Lima; International Potato Centre, Lima; International Potato Centre, Huancayo Research Station (2003)
- Working visit national agricultural research organizations of Uganda and Rwanda; NARO Kachwekano ARDC, Uganda; Institut des Sciences Agronomique de Rwanda (ISAR) (2004)
- Working visit CIP regional research station Quito, Ecuador; International Potato Centre (2006)

Invited review of (unpublished) journal manuscript (2 ECTS)
- Evaluation and Program Planning: enabling innovation in organic agriculture (2008)
- America Journal for Potato Research: trends in China's potato and sweet potato market (2009)

Competence strengthening / skills courses (2.8 ECTS)
- Planning and writing quality project proposals; ASARECA-CIAT, Kampala (2006)
- Writing successful project proposals; ASARECA (2005)

PE&RC Annual meetings, seminars and the PE&RC weekend (1.2 ECTS)
- PE&RC Weekend (2008)
- Symposium 'Farming Futures in Sub-Sahara Africa' (2008)

Discussion groups / local seminars / other scientific meetings (9 ECTS)
- Potato late blight working group for Eastern and Central Africa (2004-2007)
- UPWARD (Users' Perspectives With Agricultural Research and Development); Hanoi, Vietnam (2005)
- PRAPACE (Regional Potato and Sweetpotato Improvement Network in Eastern and Central Africa) steering committee meetings (2007)
- Strategic planning meeting of the Kenya Agricultural Research Institute (2007)
- Expert meeting on seed potato system improvement, Wageningen, the Netherlands (2009)

International symposia, workshops and conferences (10.6 ECTS)
- ISTRC-AB Meeting; Mombasa, Kenya (2005)
- 5th Bacterial Wilt conference; York, England (2006)
- Innovation for Africa symposium; Kampala, Uganda (2007)
- African Potato Association meeting; Alexandria, Egypt (2007)
- FAO Expert meeting on potato value chains; Rome, Italy (2008)
- GFRAS-CTA Conference on agricultural extension; Nairobi, Kenya (2011)

Lecturing / supervision of practical's / tutorials; 125 days
- Guest lecture in the "seed potato technology, certification and supply systems"; WUR-CDI (2008-2011)
- Developed and delivered training day on "Rapid needs and opportunity assessment" for Agri-profocus Kenya hub (2010)
- Development of a comprehensive training program on positive seed potato selection, including training of trainers and farmer group training (2005-2007)

Supervision of 4 MSc students; 60 days
- IPM of Potato bacterial wilt; Thomas Kwambai, Moi University, Eldoret, Kenya
- IPM of potato late blight; Grace Ngatia, Jomo Kenyatta University of Agriculture and Technology, Thika, Kenya
- Positive seed potato selection; Pauline Mundia, Jomo Kenyatta University of Agriculture and Technology, Thika, Kenya
- Alternative methods of breaking seed potato dormancy; Mwalimu Menza, Nairobi University, Nairobi, Kenya